Lecture Notes
in Business Information Processing 403

Series Editors

Wil van der Aalst ⓘ
RWTH Aachen University, Aachen, Germany

John Mylopoulos ⓘ
University of Trento, Trento, Italy

Michael Rosemann ⓘ
Queensland University of Technology, Brisbane, QLD, Australia

Michael J. Shaw
University of Illinois, Urbana-Champaign, IL, USA

Clemens Szyperski
Microsoft Research, Redmond, WA, USA

More information about this series at http://www.springer.com/series/7911

Karl R. Lang · Jennifer Xu · Bin Zhu · Xiao Liu ·
Michael J. Shaw · Han Zhang · Ming Fan (Eds.)

Smart Business: Technology and Data Enabled Innovative Business Models and Practices

18th Workshop on e-Business, WeB 2019
Munich, Germany, December 14, 2019
Revised Selected Papers

Springer

Editors
Karl R. Lang
Baruch College
New York, NY, USA

Bin Zhu
Oregon State University
Corvallis, OR, USA

Michael J. Shaw
University of Illinois
Champaign, IL, USA

Ming Fan 🆔
University of Washington
Seattle, WA, USA

Jennifer Xu 🆔
Bentley University
Waltham, MA, USA

Xiao Liu
University of Utah
Salt Lake City, UT, USA

Han Zhang
Georgia Institute of Technology
Atlanta, GA, USA

ISSN 1865-1348   ISSN 1865-1356   (electronic)
Lecture Notes in Business Information Processing
ISBN 978-3-030-67780-0   ISBN 978-3-030-67781-7   (eBook)
https://doi.org/10.1007/978-3-030-67781-7

This Springer imprint is published by the registered company Springer Nature Switzerland AG
The registered company address is: Gewerbestrasse 11, 6330 Cham, Switzerland

Preface

The Workshop on e-Business (WeB) is a premier annual conference on electronic business. The purpose of WeB is to provide a forum for researchers and practitioners to discuss findings, novel ideas, and lessons learned, to address major challenges, and to map out the future directions for e-Business. Since 2000, WeB has attracted valuable, novel research that addresses both the technical and organizational aspects of e-business. The 18th Annual Workshop on e-Business (WeB 2019) was held at the Munich Business School in Munich, Germany, on December 14, 2019.

"Smart Business: Technology and Data Enabled Innovative Business Models and Practices" was the theme of the WeB 2019 workshop. As the pace of technology-enabled business innovations continues to accelerate, emerging technologies and new business models have not only transformed traditional e-Business firms and markets, but have also changed both the roles of stakeholders and the connections in business networks. New technologies such as business analytics, machine learning, and artificial intelligence have been successfully applied to enhance the efficiency of e-Business transactions and the effectiveness of e-Business service offerings as well as to facilitate faster and better organizational decision making. Disruptive technologies such as blockchain and videoconferencing (e.g., Zoom, Microsoft Teams, or Webex) are fundamentally changing the way in which business is being conducted, moving further towards automated, distributed, and smarter business arrangements that are increasingly conducted online in virtual settings. At the same time, new organizational market forms like the sharing economy are transforming entire industries at a global scale. All these transformational technological and economic forces, in turn, are also re-shaping the ecosystem of e-Business

WeB 2019 provided an opportunity for academic scholars and practitioners from around the world to exchange ideas and share research findings related to these topics. The articles presented at the workshop cover a broad range of issues from the perspectives of consumers, technology users, business and management, industries, and governments using multiple perspectives, including technical, managerial, economic, and strategic thinking viewpoints. They employ various business research methods such as surveys, analytical modeling, experiments, computational models, data science, and design science.

Among the 42 papers presented at WeB 2019, we selected 20 of them through a peer-review process for publication in this LNBIP volume. We are grateful to all the reviewers for providing insightful feedback to the authors and completing their review assignments on time despite tight deadlines. And, of course, special thanks to the authors for their contributions.

December, 2020

Karl R. Lang
Jennifer Xu
Bin Zhu
Xiao Liu
Michael J. Shaw
Han Zhang
Ming Fan

Organization

Honorary Chairs

Hsinchun Chen   University of Arizona, USA
Andrew B. Whinston   University of Texas at Austin, USA

Conference Chair

Michael J. Shaw   University of Illinois at Urbana-Champaign, USA

Organizing Co-chairs

Karl R. Lang   City University of New York, USA
Bin Zhu   Oregon State University, USA
Jennifer Xu   Bentley University, USA
Xiao Liu   University of Utah, USA
Ming Fan   University of Washington, USA
Han Zhang   Georgia Institute of Technology, USA

Local Organizing Committee Chairs

Antonia Köster   University of Potsdam, Germany
Heiko Seif   Munich Business School, Germany

Program Committee

Reza Alibakhshi   HEC Paris, France
Hsin-Lu Chang   National Chengchi University, Taiwan
Michael Chau   University of Hong Kong, Hong Kong SAR
Cheng Chen   University of Illinois at Chicago, USA
Ching-Chin Chern   National Taiwan University, Taiwan
Muller Cheung   Hong Kong University of Science and Technology, Hong Kong SAR
Huihui Chi   ESCP Europe at Paris, France
Honghui Deng   University of Nevada, Las Vegas, USA
Aidan Duane   Waterford Institute of Technology, Ireland
Samuel Fosso   Toulouse Business School, France
Wencui Han   University of Illinois at Urbana-Champaign, USA
Lin Hao   University of Washington at Seattle, USA
Yuheng Hu   University of Illinois at Urbana-Champaign, USA
Jinghua Huang   Tsinghua University, China
Seongmin Jeon   Gachon University, South Korea

Contents

Crowdfunding and Blockchain

How Social Networks Dynamics can Affect Collaborative Decision Making on Crowdfunding Platforms

Yanni Hu$^{(\boxtimes)}$ (iD) and Karl Lang$^{(\boxtimes)}$

Baruch College, 55 Lexington Ave., New York, NY 10010, USA
{yanni.hu,karl.lang}@baruch.cuny.edu

Abstract. Despite the increasing phenomena that social interactions among contributors by emerging technologies influence crowdfunding decision making, little is known about how social network dynamics formed by these social interactions affect contributors' decision making. Drawing on a data set collected from an economic experiment conducted on Amazon Mechanical Turk (MTurk), we use a social network approach to investigate the effects of social network structure on collaborative decision making under a crowdfunding setting. Comparing four standard network structures – null, star, weak ties, mesh - Our analysis shows that the mesh network yields the best group collaboration performance, with social information displayed. The result of this research provides a specific and nuanced angle of the importance of social networks in emerging technology – enabled online crowdfunding.

Keywords: Crowdfunding decision making · Social network structure · Social information

1 Introduction

Crowdfunding is collaborative work - a group of people make mutual effort in reaching a fundraising goal. Contributors or backers at crowdfunding platforms always refer to each other's crowdfunding decisions when they make their own crowdfunding decisions. The Internet has offered entrepreneurs and contributors a new interaction and influential channel to support projects by social network and social information sharing Recent empirical studies have investigated the impact of crowdfunding platform design factors on contribution behavior [27], but the effects of social network on contributors' behavior have received less attention. Social network structure is of special relevance with backers' decision making if we interpret crowdfunding platforms as networks of interactions among backers and project creators. For instance, many crowdfunding sites have facilitated social media tools (Facebook and Twitter sharing) that promote social interactions. Through these interactions, contributors at crowdfunding platforms can refer to, learn from, and cooperate with each other's contribution decision making to reach the fundraising goal. As a crowdfunding project is a strategic campaign that requires

© Springer Nature Switzerland AG 2020
K. R. Lang et al. (Eds.): WeB 2019, LNBIP 403, pp. 3–17, 2020.
https://doi.org/10.1007/978-3-030-67781-7_1

backers' mutual support and collaboration to reach a fundraising goal, it is essential to understand how different network structures help with social interaction that may influence contributors' decisions to reach the fundraising goal.

Social network structure is defined as the presence of regular relationship patterns within a single social network [25, 26]. Under online settings, social network structure typically applies to the pattern in which people interact with people by many IT artifacts. Previous literature from economics, finance, and information systems recognize social network structure relates to behaviors and well-being of people in a society. Many view social network structures as information-sharing channel which influence the economic preferences and consequences [5, 17, 20]. For instance, Allen et al. [1] examine the effects of social interaction on P2P lending and find socially connected areas with more Facebook's friendship linkages have more lending activities. In crowdfunding related settings, Thies et al. [24] indicate that social networks through social media such as Facebook sharing have positive effect on backers' funding decision. Suri and Watts [23] studied network structures on cooperation behavior and find that people conditionally cooperate in response to their neighbor's decision. Fowler and Christakis [14] implements s series of one-shot public goods experiments and find that cooperation can cascade across three degrees of separation in a network. These findings provide primary motivation for this study. However, a more nuanced investigation of how individuals react to different social networks via social interaction and thus cooperate with each other during the crowdfunding is needed.

Motivated by the practical but complex effect of social network structure on crowdfunding behaviors, in this research we examine typical social network typologies on crowdfunding contribution behavior and believe that this assessment is crucial in order to understand whether and which network structure affect contributions behavior most. In addition, there is a clear need to examine how cooperation and collaboration behavior involve and evolve in different network structures as information dissemination channels. Thus, this study attempts to answer the following research question:

How does social network structure affect contributors group collaboration performance on a crowdfunding platforms?

We design an experimental crowdfunding platform deployed on Amazon Turk that enables us to manipulate social network structure and configure four standard types with different degrees of connectedness: null, weak-tie, star, and mesh network (see Table 1). We also manipulate a second variable of interest, social information, which refers to participant-specific game information shared in the network. We conduct a series of experiments on Amazon Mechanical Turk (MTurk) by inviting online individuals play a fundraising game arranged on 4 typical typologies of social networks. Data were collected from MTurk workers as participants and analyzed with one-way ANOVA and regression methodology.

Table 1. Network structure typology and centrality measures

Network structure typology and centrality measures						
Networks in 7 points		**Null**	**Weak-tie**	**Star**	**Mesh**	
	Measure:					
Community and Social Identity	Closeness Centrality: CC' (Pk)	***** ** (Inva-lid)	0.10, 0.09, 0.09, 0.07, 0.07, 0.07, 0.07 (Average: 0.08)	0.17,0.09,0.09, 0.09,0.09,0.09, 0.09 (Average: 0.10)	0.17,0.17,0.17, 0.17, 0.17,0.17, 0.17 (Average: 0.17)	
Social Identity		Null	< Weak tie	< Star	< Mesh	
Connectedness and So-cial Dis-tance	Betweeness Centrality: CB' (Pk)	***** ** (Inva-lid)	9,8,8,0,0,0,0 (Average: 3.57)	15, 0,0,0,0,0,0 (Average: 2.14)	0,0,0,0,0,0,0 (Average: 0)	
Social Distance		Mesh	< Weak tie <	Star	< Null	

2 Theory and Hypotheses Development

2.1 Social Network Typologies

Many previous literature in crowdfunding and micro-financing have different concep-tualizations for social network dynamics. For example, Hong et al. [16] have placed "network embeddedness" as a characteristic of social network that has an important role in influencing crowdfunding campaigns via social media. They acknowledge that the nature of embeddedness in social networks lead to greater social influence and higher social information sharing. Among the sociological and economic literature, the most popular view is "social capital" view that treat social ties (especially weak ties) as pool of social assets or resources embedded in social networks [3, 19, 21]. Other research conceptualizes social network dynamics as social connections that can improve people's financial performance in group lending [18]. These research measure cultural and (or) geographic proximity as a proxy of social connection between each pair of individu-als. Of all the previous studies, the major concern is whether social network dynamics formed by social interactions significantly impact users' information intake, evaluation, and decision making. We exploit and discuss this question under a crowdfunding setting in this study, since such an assessment is crucial in understanding the role of social network on contribution performance, as well as helping with design of crowdfunding platforms to reap the benefits of social networks.

We design four typical types of social network structures that aligned with the typolo-gies conceptualized by Freeman [15] (See Table 1), and identify these structures with social distance theory and social identity theory in an online crowdfunding setting. The reason we select and explore these four typologies is that they each represent a type of organization of structure that could be synthesized and compare with each other by their different structural properties. In particular, we use closeness and betweenness centrality measures theorized by Freeman to categorize these four typologies and theorize our hypotheses. Closeness centrality measures one's closeness to the others, and it indicates how "close" one's relationship with others based on a calculation of reciprocal of sum of the shortest path between him or herself as a node in a graph and all other nodes. It also can be regarded as a measure of how fast it takes to spread information from a node to the others. We can observe that the Mesh network yields the highest closeness between its all pairs of nodes, than the Star, Weak tie, and Null. Betweenness centrality is another measure of centrality and is calculated by the number of times a node that lies on the shortest path between other nodes. The higher the betweenness, the lower the social distance score for strategically located people. Therefore, the Mesh yields the shortest social distance on average than the Star, Weak tie, and the Null, by lowest betweenness centrality. We rank each type of networks with the aforementioned closeness centrality and betweenness measures, and theorize our hypotheses using social identity theory and social distance theory.

We manipulate our experiment treatments based on the four types of structures in an online fundraising setting that a fundraising project solicits donation from an online community that consists of 7 members. The null structure is designed as a pattern where members as nodes do not establish any relationship with the others. In the star structure, however, the central node can represent a community leader connecting to the other nodes. Members independently communicate to the leader who is assumed to be more structurally central than others. The weak tie structure represents two loosely connected groups linked by a leader (or influencer) who is supposed to transfer information across the two groups. Finally, in the mesh structure, every member can communicate directly with each other, but there is no essential "leader" role in this structure, representing a democratic community.

Previous studies have cited links between the structural properties of social network and sharing and creative behavior in organizational context [2, 13]. Current literatures in crowdfunding suspect structural properties of a social network may influence decision making. For example, Fowler and Christakis [14] predicts social networks influence the evolution of contribution behavior by helping spread the cooperation norm across individuals in a network. Thus, we infer that social network links and ties in social network structure help bridge individuals and information gap by promoting information sharing, which will influence people's decision making in crowdfunding. Thus, we expect

H1: Overall, Social network structure matter for crowdfunding decision making, measured by group collaboration performance.

2.2 Community and Social Identity Theory

We define community as a property of such networks that are highly compact and almost everyone in the networks has relational ties with each other. Social identity theory has defined social identity as an individual's perception about his or her membership and belonging to the social group or community We relate community with social with social identity since the sentiment of social identity tends to depersonalize the individual but contribute to the development of group or community attachment and community success especially as one develops self-categorization and collective-identification that relate to perceived similarities among group members [10, 22]. For the mesh network structure, it possesses the highest graph closeness centrality since it has most links and bridges between all individual which promote communication and perceived similarity among group members, we infer the mesh network yields the highest group collaboration performance by linking and bridging individuals to efficiently cooperate with each other. Thus we expect

Hypothesis 2: *Degree of social identity is positively associated with group collaboration performance.*
Hypothesis 2a: Degree of social identity is positively associated with total group contribution.
Hypothesis 2b: Degree of social identity is negatively associated with distance to threshold equilibrium.
Hypothesis 2c: Degree of social identity is positively associated with success rate.

2.3 Connectedness and Social Distance Theory

Social connectedness by previous literature has been defined as intensity of friendship links, in particular, by geographic distance [4]. Social connectedness is of great relevance with users and group decision making if we explain it from the impact of social distance theory. Previous studies [11, 12] have conceptualized three dimensions of social distance: affective social distance, normative social distance, and interactive social distance. Here, we take interactive social distance which focuses on the frequency and intensity of interaction facilitate social distance of two social groups. By communicating and interacting with each other, interpersonal bonds arise and social distance between two social groups decrease [22]. For instance, researchers found that entrepreneurs who build a network of direct ties could shorten the social distance, induce a trustful relationship in which both parties are motivated to maintain, and generate a sense of obligation and cooperation between investors and them. Previous literature also find that decrease of social distance promotes social interactions and empathy among individuals which result in higher levels of cooperation [6]. That is to say, when individuals' social distance decreases, "others" are not just some unknown ones but ones deserve other-regarding behaviors. Because the mesh network gains the lowest social distance by owning highest number of strategically located people that result in lowest betweenness centrality, than the star, weak tie and the null, we expect

Hypothesis 3: *The lower the social distance is, the higher the group collaboration performance yields. Therefore, degree of social distance is negatively associated with group collaboration performance.*
Hypothesis 3a: Degree of social distance is negatively associated with total group contribution.

Hypothesis 3b: Degree of social distance is positively associated with distance to threshold equilibrium.

Hypothesis 3c: Degree of social distance is negatively associated with success rate

2.4 Group Collaboration Performance

Our theoretical concern is group collaborative decision making by group performance in this study. In this study, we define collaboration as a cooperation process in which individuals interact, share information, and make mutual effort to reach the fundraising goal. We measure group collaboration performance by 1) group total contribution 2) distance to threshold equilibrium 3) Success and failure rate.

These three dependent variables each points to a different but important angle of measurement of group collaboration performance. Group total contribution is the indicator of generosity of contributors [8, 9] The second measure distance to threshold equilibrium is calculated by the absolute value between group total contribution and the economically most efficient outcome (threshold value 17.5 points). A shorter absolute distance to threshold indicates the group coordinated closely around the threshold, which shows efficiency of collaboration regardless of whether group total contribution is surpass or below the fundraising threshold. Finally, the third measure, success and failure rate, calculated by rounds that successfully reach the funding goal, indicates the principle success of the collaboration (See Fig. 1).

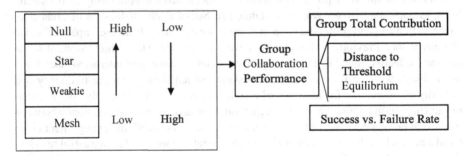

Fig. 1. Research model

3 Research Methodology

3.1 Experimental Design

We implement our fundraising game by adapting a widely used public goods game from the economics literature [8]. It allows us to theoretically examine optimal, rational collaborative decision making behavior. Our use of this fundraising game is aimed to present the participants with non-trivial tradeoff decisions. The rational is that if the group reaches the fundraising goal it will suffice to implement the proposed project. Hence, there is the question of how much should the individual group member contribute, and what is too little and what would be too much to efficiently reach the goal? How does the group respond to possible free-riding behavior? To what extend does the group learn to collaborate better over the course of multiple, repeated rounds of the experiment. This represents a non-trivial collaborative decision making problem.

The experiment is designed with a financial incentive to induce rational decision-making behavior. If the group contribution does not reach the fundraising goal, no reward will is be given to the contributors But if it succeed, every contributor individually receives a fixed reward for successfully supporting the project. In either case, the contribution amount will not be returned. Our model is different from all-or-nothing mechanism in some crowdfunding sites such as Kickstarter in that we don't give back contribution if members do not reach the fundraising goal. We design our experiment mechanism as such because we need to induce a non-trivial decision for each participant to either cooperate to contribute to the goal, or contribute an amount cautiously since contributions out will never be returned. Through this way, we can examine group collaboration behavior more directly.

During the fundraising campaign, the stated funding goal is 17.5 game dollars. We convert the threshold 17.5 from Cadsby and Maynes paper based on our group size of 7. Each session involves 7 individuals who are each provided with an endowment 10 game dollars. Each participant must privately decide how much to contribute, C_i, $C_i \in [0, 10]$ where $i \in [1,..., N]$ to the fundraising project. If the threshold 17.5 is reached (project succeed), each participant will receive a reward of 5 game dollars; If not (project failed), each participant receive nothing. After each round, participants are displayed with others contribution amount based on the social network assigned to them, as well as their own earnings so that they have a clue of how many to contribute in next round. The game is played repeatedly 10 rounds. The points the participants earn in each round will be converted to U.S. dollars at the end of the game, according to the conversion rate: 1 point = 6 cents. Each participant is able to earn around $3 plain fee plus an average of $5–$10 performance-based payment, which was paid out after the game. (See feuqations below and Table 2. Experimental Parameters). Ui (individual's payoff) =

$$10 - C_i, \text{ if } \Sigma_{i=1}^{N} = C_i < 17.5 \tag{1}$$

$$10 - C_i + 5, \text{ otherwise} \tag{2}$$

Table 2. Experiment parameters

Group size:	7
Endowment	10
Threshold level	17.5
Reward level	5
Rounds	10 rounds
Conversion	1 = 0.06 US dollars

3.2 Procedures

We hire participants through Amazon Mechanical Turk (MTurk) where a flat participation fee plus performance payment are paid to the platform workers. We use MTurk
workers as our participant pool because MTurk is a suitable platform for interactive
experiments where participants need to wait for others to form a group, exchange information, and make joint decisions. MTurk workers are recruited through posting of HIT
on the MTurk platform where workers can freely accept the HIT. We manipulate one
treatment variables, social network structure at four levels (null, star, weak-tie, mesh).
Conceptually, we model also social information facilitation but keep it fixed at one level
in the current study (social information present). We separately measure three dependent
variables (total group contribution, distance to equilibrium, success rate). The experiment was conducted with a total of 12 sessions, with 3 sessions for each of the 4 network
structures. A total of 84 participants are recruited. The subjects are organized into a
group of size 7 per session to play the fundraising campaign.

We manipulate online social network structures by social profile sharing and at the
same time displaying social information (others' contribution amount). At the beginning
of the experiment, we induce the participants' perception of social interaction by asking
participants to answer a series of Facebook-like questions and complete their social
profile (see Fig. 2). Then, to induce a sense of social interaction to other members in the
group, participants as community members will share their social profile based on the
social network structure they stay in. For instance, in the star network, User 4 as a regular
member can only see User 1 (influencer)'s contribution amount while the influencer can
see all others' contributions (see Fig. 3). These two features together allow us to draw
social interactions between participants by knowing and learning from others' social
information and contribution amount. These two features are automatically created for
the participants when the experiment begins based on the given social network structure
in the particular treatment. The profile sharing is done automatically done by the system
after participants complete the profile creation. The profiles are only shared along the
direct connections in the given network structure. In the null structure, no profiles are
shared, while everyone gets to see everyone else's profiles in the mesh structure. In the
star structure, everyone gets to see the leader's profile but none of the others, while
the leader sees everyone's profiles. Similarly, in the weak-tie structure, the influencer
(leader) sees profiles from both subgroups, while the peripheral members only see some
profiles from members within their subgroup.

4 Data Analysis and Preliminary Findings

4.1 Descriptive Analysis and Group Homogeneity

In total, 12 sessions (groups) of the experiment were conducted with 3 sessions (groups)
run for each type of the social network structure. A total of 84 participants Data were
collected from each group over 10 repeated rounds of the experiment. As one of the
assumptions for running ANOVA test, group homogeneity should be tested. We ran
Levene's test and get $p = 0.043$ which is close to the recommend $p = 0.05$ value that
is satisfying ANOVA test group equal variance assumption (See Table 3). We proceed

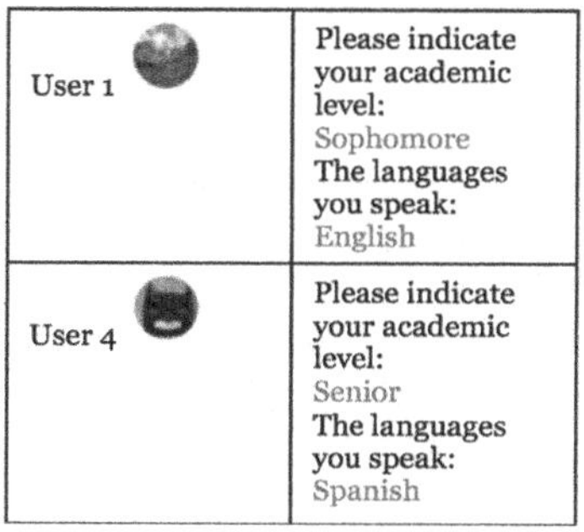

User 1	Please indicate your academic level: Sophomore The languages you speak: English
User 4	Please indicate your academic level: Senior The languages you speak: Spanish

Fig. 2. User 1 screen for social profile sharing process in the star network

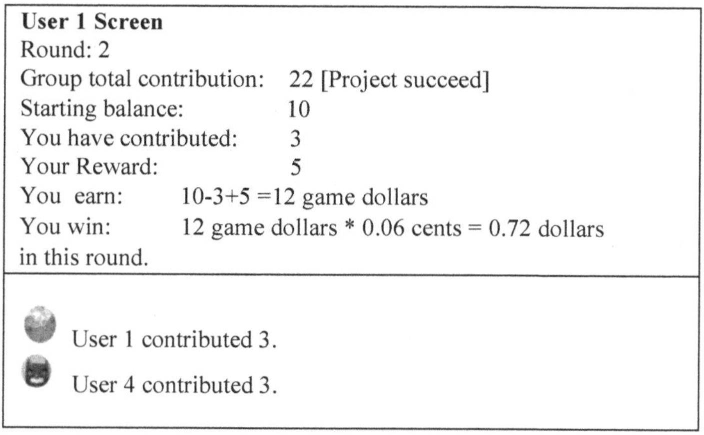

User 1 Screen
Round: 2
Group total contribution: 22 [Project succeed]
Starting balance: 10
You have contributed: 3
Your Reward: 5
You earn: 10-3+5 =12 game dollars
You win: 12 game dollars * 0.06 cents = 0.72 dollars
in this round.

User 1 contributed 3.

User 4 contributed 3.

Fig. 3. User 1 screen for contributing screen in the star network

to run an ANOVA F test to test whether the 4 network structures differ significantly in influencing group collaboration performance by group total contribution.

Table 3. Group homogeneity test

Levene statistic		df1	df2	Sig.
Based on mean	2.80	3	126	.043
Based on median	2.72	3	126	.047
Based on median and with adjusted df	2.72	3	115.13	.048
Based on trimmed mean	2.81	3	126	.042

Table 4. Descriptive for group total contribution

N		Mean	Std. deviation	Std. error	Minimum	Maximum
Null	30	26.00	4.71	.86	15.00	35.00
WeakTie	30	25.10	3.74	.68	15.00	33.00
Star	30	23.97	5.75	1.05	15.00	37.00
Mesh	30	22.40	4.15	.76	17.00	33.00
Total	120	24.37	4.78	.44	15.00	37.00

4.2 Group Total Contribution

In total, 12 sessions (groups) of the experiment were conducted. ANOVA test (see Table 5) shows that the 4 social network structure yields significantly different group total contribution with a p-value of 0.007. Group members contribute most in the null (mean = 26.00) and least in the mesh network (mean = 22.40), which indicates a decreasing trend due to the increased interaction of the members within a network structure (see Table 4). This result support hypothesis 1 that group collaboration performance by total contribution differ significantly across four structures. However, contributing more than the threshold is inefficient as it will not bring more reward to the members and may cause a waste of social welfare. Therefore, we present the second dependent variable, distance to threshold equilibrium, on group collaboration performance.

Table 5. ANOVA test of group total contribution variance

Group total contribution					
Sum of squares		df	Mean square	F	Sig.
Between groups	260.03	3	86.68	4.206	.007
Within groups	2596.77	126	20.61		
Total	2856.80	129			

4.3 Distance to Threshold Equilibrium

Figure 4 displays group total contribution by each of the four networks, and Fig. 5. illustrates the averaged 10 rounds distance to threshold equilibrium by the four network structures. Group members in the mesh gets the closest to the threshold equilibrium (mean = 4.90) than the other network structures (Star = 6.47, Weak tie = 7.60, Null

$= 8.50$), which indicate the mesh groups collaborate most efficiently by getting close to the fundraising goal. The ANOVA test displays significant differences among the four structures with $F = 6.185$ and $p = 0.001$. This suggests averaged distance to threshold equilibrium across four network structures differ significantly. Overall, the mesh network yields the best group collaboration performance by collaboration efficiency, following by the star, weak tie, and the null. The result supports hypothesis 2a, 2b and 3a, 3b that lower social distance and higher social identity among members help passing information across individuals in a network and bring high group cooperation via low distance to threshold equilibrium. In a mesh network, individuals are more likely to see and evaluate the others' contribution amount then make their cooperative or non-cooperative decision.

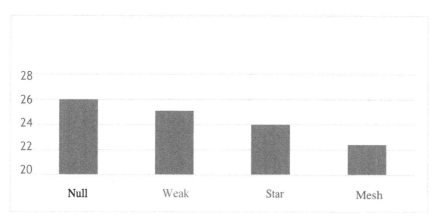

Fig. 4. Group total contribution by 4 social networks

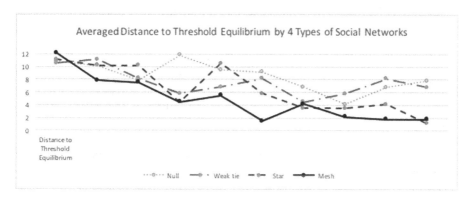

Fig. 5. Distance to threshold by 4 social networks

Because we would also like to know how often the project was successfully funded over the 10 rounds, we count the number of rounds that reached the group contribution threshold level. We can observe that the star groups yield the lowest success rate, which suggest a lower cooperation and collaboration accuracy in achieving the threshold equilibrium − 17.5 point, than the other 3 groups. The results are not quite supporting our hypothesis H2c and H3c.

Our analysis indicates a wider standard deviation of contributions among the star group members. The star network limited connections between members by only depicting leader's decision to each of the other group members. We may infer that A dyad relationship may not act as efficiently as a triangle to transmit useful information among the members. Members could not observe how many points others have contributed but only the leader's, thus leading to a deficiency in calculating themselves contribution points to efficiently coordinate around the threshold but only following the leader's decision making. Table 6 presents the success rate statistics for each of 4 network structures.

Table 6. Averaged distance to equilibrium and public goods project success rate

Social network structures	Mesh	Star	Weak tie	Null
Success rate: number of rounds project was funded	29/30	27/30	29/30	29/30
Averaged group distance to threshold equilibrium	4.90	6.47	7.60	8.50

We conduct regressions (Table 7) for average group total contribution in 10 rounds (first column), average group total contribution in last 5 rounds (second column), and average distance to threshold (third column). We make first column dummy variables and set the mesh structure as baseline, which means that group contribution in the star network is 1.567 game dollar higher than the mesh; group contribution in the weak-tie network is 2.700 game dollar higher than the mesh; and finally, group contribution in the null network is 3.600 game dollar higher than the mesh. Since extra amount of group contribution can cause waste of resource which cannot be used into the crowdfunding project, we conclude that group members in the mesh network achieve the best and most efficient collaboration performance, since their members cooperate most closely around the project threshold of 17.5 by arriving a group contribution at 22.4 game dollars. All of the social networks yield significant effect on group collaboration performance, except the star network ($p = 0.194$). Interestingly, the last 5 rounds average results are interesting because they indicate a higher R square (0.261) that may show greater social network's effect on group collaboration performance, suggesting a propensity that members collaboration ability and learning effect evolve more quickly and significantly in the last 5 rounds of the fundraising campaign. We involve distance to threshold as another important dependent variable. Regression to distance to threshold yield same coefficient and p-value with those for the GT (group total contribution) because in our data, all groups contributed over the threshold, which means no groups made a group contribution lower than 8.75 (a threshold under which free-riding tendency is determined). However, we include this dependent variable since it is another essential measure of group collaboration performance that is distinct with group total contribution. A higher group total contribution indicates generosity of group members but not necessary the efficiency of their collaboration; instead, closeness to threshold equilibrium indicates group collaboration efficiency-how close the members contribute around the threshold. Finally, we also examine the success and failure, since collaboration efficiency by distance to threshold can be either positive or negative. A negative number means a group contributes under threshold and project fails to get enough fund. From Table 7 we can conclude that the mesh still yields the best collaboration performance. However, the star network is unstable since it yields that lowest success rate.

Table 7. Regression table (coefficient, t - test statistics and p - values)

Independent variables	Dependent variables		
	GT (10 rounds)	GT (last 5 rounds)	DTT
Constant	22.400 (26.392) (0.000)	19.800 (22.100) (0.000)	4.900 (5.773) (0.000)
Null (compare to Mesh)	3.600 (2.999) (0.003)	4.667 (3.683) (0.001)	3.600 (2.999) (0.003)
Weak tie (compare to Mesh)	2.700 (2.249) (0.026)	4.400 (3.473) (0.001)	2.700 (2.249) (0.026)
Star (compare to Mesh)	1.567 (1.305) (0.194)	1.333 (1.052) (0.297)	1.567 (1.305) (0.194)
Adjusted R square	0.080	0.261	0.080

5   Conclusion

In this study we provide strong experimental evidence that social network structure with sufficient social information does matter to group collaborative decision making in a fundraising campaigns. Our experiment has offered us a unique opportunity to directly observe the influence of social network dynamics on the contribution behavior in crowdfunding through MTurk experiments. Group in mesh network yields the better collaboration performance than the other three social network structures by passing social information most efficiently through the most connected channels. People in the mesh network cooperate most closely around fundraising goal. This suggests that people exposed to a highly - connected social network where everybody is linked with each other tend to coordinate and collaborate most efficiently around the threshold. From a social psychology perspective, highly connected social network may escalate the helping and social learning behavior that each member accommodates behavior by seeing and knowing what others are doing through social interaction during profile sharing and contribution procedures.

The investigation as well as the results of this study have important implications for project initiator or promoter as well as crowdfunding sites design in IT-enabled environments. Crowdfunding sites as well as project initiator might seek intentionally to leverage social interactions to nudge contributors to form more online social relationships. In addition, showing sufficient social information, such as crowdfunding goal and other members' contribution amount, provide especially important background for online contributors to infer and adjust their decision to reach the crowdfunding goal.

One limitation of this preliminary study deals with the typical but small network typology, which may limit the generalizability of the findings. Future research could investigate more topologies of structures by involving larger and more complex networks in field experiment to examine whether the results still hold. Our research takes a first step in understanding the domain of social network structure with social information on crowdfunding contribution decision making. We hope our analysis provide initial evidence of impact of social network dynamics on crowdfunding decision making, which serves reasonable recommendations to the crowdfunding platform as well as project initiators.

References

1. Allen, L., Peng, L., Shan, Y.: Social interactions and peer-to-peer lending decisions. Hong Kong University FinTech Conference (2019)
2. Amabile, T.M., Conti, R., Coon, H., Lazenby, J., Herron, M.: Assessing the work environment for creativity. Acad. Manage. J. **39**, 1154–1184 (1996). https://doi.org/10.2037/256995
3. Angelusz, R., Tardos, R.: The strength and weakness of weak ties. In: Values, Networks and Cultural Reproduction in Hungary, pp. 7–23 (1991)
4. Bailey, M., Cao, R., Kuchler, T., Stroebel, J., Wong, A.: Social connectedness: measurement, determinants, and effects. J. Econ. Perspect. **32**, 259–280 (2018). https://doi.org/10.1257/jep.32.3.259
5. Berger, K., Klier, J., Klier, M., Probst, F.: A review of information systems research on online social networks. Commun. Assoc. Inf. Syst. **35**, 145–172 (2014). https://doi.org/10.17705/1CAIS.03508
6. Bohnet, I., Frey, B.S.: Social distance and other-regarding behavior in dictator games: comment. Am. Econ. Rev. **89**, 335–339 (1999). https://doi.org/10.1257/aer.89.1.335
7. Burt, R.S.: The network structure of social capital. Res. Organ. Behav. **22**, 345–423 (2000). https://doi.org/10.1016/S0191-3085(00)22009-1
8. Cadsby, C., Maynes, E.: Gender and free riding in a threshold public goods game: experimental evidence. J. Econ. Behav. Organ. **34**, 603–620 (1996). https://doi.org/10.1016/S0167-2681(97)00010-3
9. Cadsby, C., Maynes, E.: Choosing between a socially efficient and free-riding equilibrium: nurses versus economics and business students. J. Econ. Behav. Organ. **37**, 183–192 (1998). https://doi.org/10.1016/S0167-2681(98)00083-3
10. Code, J., Zaparyniuk, N.: Social identities, group formation, and the analysis of online Communities. In: Handbook of Research on Social Software and Developing Community Ontologies, IGI Global (2010)
11. Eveland Jr., W.P., Nathanson, A.I., Detenber, B.H., McLeod, D.M.: Rethinking the social distance corollary: perceived likelihood of exposure and the third- person perception. Commun. Res. **26**, 275–302 (1999). https://doi.org/10.1177/009365099026003001

12. Fiedler, F.E.: The psychological-distance dimension in interpersonal relations. J. Pers. **22**, 142–150 (1953). https://doi.org/10.1111/j.1467-6494.1953.tb01803.x
13. Ford, C.M.: A theory of individual creative action in multiple social domains. Acad. Manage. Rev. **21**, 1112–1142 (1996). https://doi.org/10.2307/259166
14. Fowler, J.H., Christakis, N.A.: Cooperative behavior cascades in human social networks. Proc. Natl. Acad. Sci. **107**, 5334–5538 (2010). https://doi.org/10.1073/pnas.0913149107
15. Freeman, L.C.: Centrality in social networks, conceptual clarification. Soc. Netw. **79**, 215–239 (1979). https://doi.org/10.1016/0378-8733(78)90021-7
16. Hong, Y., Hu, Y., Burtch, G.: Embeddedness, prosociality, and social influence: evidence from online crowdfunding. MIS Q. **42**, 1211–1224 (2018). https://doi.org/10.25300/MISQ/2018/14105
17. Jackson, M.O., Rogers, B.W., Zenou, Y.: The economic consequences of social-network structure. J. Econ. Lit. **55**, 49–95 (2017). https://doi.org/10.1257/jel.20150694
18. Karlan, D.S.: Social connections and group banking. Econ. J. **117**, 52–84 (2007). https://doi.org/10.1111/j.1468-0297.2007.02015.x
19. Lin, N.: Building a network theory of social capital. Connections **22**, 28–51 (1999). https://doi.org/10.4236/ib.2011.32017
20. Mislove, A.E.: Online social networks: measurement, analysis, and applications to distributed information systems. Dissertation. Rice University (2009)
21. Postelnicu, L., Hermes, N., Szafarz, A.: Defining social collateral in microfinance group lending. Working paper. Université Libre de Bruxelles (2013)
22. Ren, Y., et al.: Building member attachment in online communities: applying theories of group identity and interpersonal bonds. MIS Q. **36**, 841–864 (2012). https://doi.org/10.2307/41703483
23. Suri, S., Watts, D.J.: Cooperation and contagion in web-based, networked public goods experiments. ACM SIGecom Exchanges **10**, 3–8 (2011). https://doi.org/10.1371/journal.pone.0016836
24. Thies, F., Wessel, M., Benlian, A.: Effects of social interaction dynamics on platform. J. Manage. Inf. Syst. **33**, 843–873 (2016). https://doi.org/10.1080/07421222.2016.1243967
25. Wasserman, S., Faust, K.: Social Network Analysis: Methods and Applications (1994)
26. Woodman, R.W., Sawyer, J.E., Griffin, R.W.: Toward a theory of organizational creativity. Acad. Manage. Rev. **18**, 293–321 (1993). https://doi.org/10.2307/258761
27. Xu, M., Cai, Y.: Crowdfunding strategically: a signaling examination on the determinants of crowdfunding success. In: The 3rd Annual Symposium on Data Analytics, Baruch College (2019)

Go in the Opposite Direction? The Impact of Unavailability on Crowdfunding Success

Wanghongyu Wei[✉] and Michael Chau

The University of Hong Kong, Hong Kong Sar, China
yhww@connect.hku.hk, mchau@business.hku.hk

Abstract. Reward-based crowdfunding are increasingly playing an important role in raising financial capital for small projects. The most important goal for creators on the platform is to successfully raise enough capital for their projects. Our study aims to provide a new angle to understand backer's decision-making process on pledging behavior by uncovering how different dimensions of unavailability influence crowdfunding success. By analyzing more than 400,000 projects on Kickstarter, we find that time-based unavailability can indeed improve the possibility of success while quantity-based unavailability has negative impacts. Besides, each dimension of unavailability can influence how individuals interpret other dimension of unavailability by changing the way how individuals receive and process persuasive information.

Keywords: Crowdfunding · Scarcity · Decision making · Heuristic-Systematic model · Information asymmetry

1 Introduction

1.1 A Subsection Sample

The rapid advancement of information technology has spawned a number of FinTech revolutions, such as crowdfunding, social investing, social lending, mobile payment, blockchain, digital currency, algorithm trading, etc. Mollick (2014) defines crowdfunding as "the efforts by entrepreneurial individuals and groups – cultural, social, and for-profit – to fund their ventures by drawing on relatively small contributions from a relatively large number of individuals using the internet, without standard financial intermediaries." Models of crowdfunding becomes stable after several years' development. Generally speaking, there are four types—donation-based model, lending-based model, reward-based model, and equity-based model—classified according to the "reward" (Belleflamme, Lambert, and Schwienbacher 2014).

This study focuses on reward-based crowdfunding, which has become a nonnegligible alternative for entrepreneurs to raise financial capital. As of August 2019, the leading reward-based crowdfunding platform in the world, Kickstarter, has already raised 4.5 billion US dollars for more than 169,000 successful projects[1]. Naturally, the determinants of success of raising capital attract oceans of attention from both practitioners and

---

[1] https://www.kickstarter.com/help/stats (accessed August 2019)

© Springer Nature Switzerland AG 2020
K. R. Lang et al. (Eds.): WeB 2019, LNBIP 403, pp. 18–28, 2020.
https://doi.org/10.1007/978-3-030-67781-7_2

researchers (Burtch, Ghose, and Wattal 2014; Hong, Hu and Burtch 2018; Li and Wang 2019; Lin, Prabhala, and Viswanathan 2013; D. Liu, Brass, Lu, and Chen 2015; Xu and Chau 2018; Younkin and Kuppuswamy 2017). Although considerable progress has been made, explanations from different theoretical perspective remain indispensable.

Our study investigates the effect of "unavailability" on project success, which is seemingly opposite to previous literature. In the context of crowdfunding, we define unavailability as the explicit constraints to back a project. To further elaborate the influence of unavailability, we do not treat unavailability as a unitary construct, but separate the unavailability construct into two different dimensions—quantity-based unavailability, and time-based unavailability. Quantity-based unavailability refers to the constraints that restrict the total number of backers who are allowed to pledge for certain reward in a project. Those rewards are backed according to the first-come-first-served rule. Time-based unavailability refers to the constraints that restrict the time period during which backers can pledge for the project.

To facilitate more precise understanding of the two definitions, we use two real projects on Kickstarter as the example, one named "The World's Warmest NECK GAITER with Stash Pocket" (noted as project A)[2] and the other named "EcoQube-Desktop Ecosystem That Grows Flowers and Herbs" (noted as project B)[3]. Project A has 8 rewards to choose and 2 of them limit the total number of backers allowed to pledge while project B has 11 rewards to choose and 4 of them limit the total number of backers allowed to pledge. Quantity-based unavailability is built on the percentage of rewards which limit the total number of backers to pledge. Thus, quantity-based unavailability should be higher for project B than project A (4/11 is larger than 2/8). A higher percentage of rewards with quantity limit of backers represents higher quantity-based unavailability.

Time-based unavailability is built on the length of funding period. Project A is allowed to pledge from Nov 12, 2015 to Dec 12, 2015, 30 days in total, while project B is allowed to pledge from Dec 1, 2013 to Jan 12, 2014, 42 days in total, so the time-based unavailability should be higher for project A than project B. Shorter funding period represents higher time-based unavailability.

Furthermore, the two dimensions do not necessarily have the same impact, or their impacts share the same underlying mechanisms, so we also investigate their interacting effect. One of the most salient characteristics of crowdfunding is that it lowers the threshold for anyone to start a project and invest in a project. However, we claim that the unavailability of the projects is indeed able to enhance the possibility of crowdfunding success. Our argument is built on commodity theory (Brock 1968) and the heuristic-systematic model explaining how individuals process persuasive information (Chaiken 1980).

Our paper has two major contributions. First, it contributes to the large stream of literature on unavailability (Cachon, Gallino, and Olivares 2018; Cui, Zhang and Bassamboo 2018; Kremer and Debo 2015; Q. Liu and Van Ryzin 2008; Lynn 1991; Stock and

---

[2] https://www.kickstarter.com/projects/1983290420/the-worlds-warmest-neck-gaiter-with-stash-pocket (accessed August 2019)

[3] https://www.kickstarter.com/projects/kevinzl/ecoqube-desktop-ecosystem-that-grow-flowers-and-he (accessed August 2019)

Balachander 2005), which demonstrates that individuals interpret different dimensions of unavailability differently. Our research extends this stream of study by identifying that individuals believe time-based unavailability to be reliable and authentic signals for internal project quality, but they consider quantity-based unavailability skeptically. Furthermore, we show that time-based unavailability and quantity-based unavailability can jointly influence crowdfunding success, indicating that time-based unavailability can influence how individuals interpret quantity-based unavailability.

Second, our study adds to the large body of work on backer's decision making in reward-based crowdfunding (Colombo, Franzoni and Rossi–Lamastra 2015; Dai and Zhang 2019; Li and Wang 2019). Backers in online crowdfunding platform exhibit different behaviors under different dimensions of unavailability. The effect of unavailability is delivered though influencing how individuals receive and process persuasive messages. Time-based unavailability induces individuals to rely on heuristic cues to infer internal quality while quantity-based unavailability is only perceived to be sales tactics. Our further analysis provides evidence that combining the two dimensions of unavailability can amplify the positive impact on crowdfunding success.

2   Literature Review and Hypothesis Development

2.1   Theoretical Background Theoretical Background

Our study is closely related to the unavailability literature which roots in commodity theory (Brock 1968). Scarcity has huge psychological power because individuals treat it as the heuristic cue for value (Cialdini 2007). Unlike previous studies which usually focus on one dimension of unavailability, we systematically investigate how each dimension of unavailability influences on crowdfunding success and their interacting effect. In addition, our study relates to the large volume of literature concerning the factors influencing investors' decision making on crowdfunding platforms. A number of factors are identified to exert significant impact, such as geographic distance (Agrawal, Catalini and Goldfarb 2015; Burtch et al. 2014; Lin and Viswanathan 2015), communication (Xu and Chau 2018), cultural differences (Burtch et al. 2014), founder's race (Younkin and Kuppuswamy 2017), and project prosociality (Dai and Zhang 2019; Li and Wang 2019). Different facets of social capital also matter, such as, friendship (Lin et al. 2013; D. Liu et al. 2015), social networks (Lukkarinen, Teich, Wallenius, & Wallenius, 2016), embeddedness (Hong et al. 2018), and reciprocity (Colombo et al. 2015).

2.2   Quantity-Based Unavailability and Crowdfunding Success

Quantity-based unavailability refers to the constraints that restrict the total number of certain rewards or products. It is generally shown to increase the perceived value (Lynn 1991) with evidence from a wide range of products as well as in a broad range of situations. These products include but are not limited to cookies (Worchel, Lee and Adewole 1975), books (Verhallen and Robben 1994), wines (Van Herpen, Pieters and Zeelenberg 2009), and automobiles (Cachon et al. 2018). Besides, the effect that quantity-based enhances value remain robust even when individuals are suffering from financial constraints (Sharma and Alter 2012). Prior researchers identify numerous mechanisms disclosing the association between quantity-based unavailability and increased value, such as the need for uniqueness (Fromkin and Snyder 1980), psychological reactance (Clee and Wicklund 1980), and naïve economic inferences (Lynn 1992).

Contrary to prior studies, we argue that prior old offline conclusion cannot be generalized to online context. The theory most cited by prior studies is come up with in 1984. At that year, less than 10% of US households had a computer and none of them had internet, not to say other countries. The shift from offline context to online context changes the cognitive and affective processes which establish the effect of offline quantity-based unavailability. In detail, the premise of quantity-based unavailability to be effective is that potential backers believe the signal of quantity-based unavailability is true. This premise is hardly held on online crowdfunding platform because of the high information asymmetry. Potential backers are naturally skeptical of costless signal of value. Unlike time-based unavailability which greatly threatens the success of crowdfunding, quantity-based unavailability does not bring such risk. There are two reasons. First, most of the projects offer very high upper bound which are seldom filled up. Second, most of the projects provide many categories of rewards. Thus, before the quantity limitation is reached, the funding goal has been reached, which indeed brings no failing risk. Frivolous overuse of quantity-based unavailability claims reinforces backers' suspicion of the authenticity of these signals. Thus, without other costly as well as reliable signal, the effect of quantity-based unavailability should be negative and we hypothesize that,

H1: Quantity-based unavailability is associated with lower possibility of funding success.

2.3 Time-Based Unavailability and Crowdfunding Success

Time-based unavailability refers to the constraints that restrict the time period when certain rewards or products can be obtained, which may motivate potential backers to pledge through three mechanisms. First, prospect theory suggests that individuals are naturally loss aversion because they have a "value function" which is positive as well as concave over gains while is negative, convex, and more steeply sloped over losses (Kahneman and Tversky 1979). Shorter funding duration increase the possibility that potential backers lose the opportunity to get rewards permanently if they do not pledge before deadline. Since individuals are more sensitive to loss, the framing of missing deadline as permanent loss increases the likelihood for individuals to pledge. Specifically, the increased likelihood should be a function of individuals framing a reward initially as a potential gain and then reframing it as a permanent loss once the potential backers process the deadline information.

Second, regret theory also sheds light on how time-based unavailability influences pledge intention. Promotions with time limitation are found to accelerate purchase more than promotions without time limitation, and besides, purchase intention is shown to increase dramatically as deadline approaches (Aggarwal and Vaidyanathan 2003; Inman and McAlister 1994). Similarly, as the deadline of a project approaches, potential backers may feel impending regret about the rewards they are losing if they do not pledge. The fact that potential backers will miss the reward forever creates the perception of time-based unavailability, and it induces pledge action by taking advantage of potential backers' fear of "missing" (Cialdini 2007). The regret effect is particularly salient in the context of crowdfunding because projects introduce new products to the market and individuals can hardly find the same product or substitute after they miss the chance to pledge.

Third, time-based unavailability motivates individuals to take risk. Under time pressure, the attractiveness of risky choice increase (Young, Goodie, Hall and Wu 2012). What's more, with shorter time period to make a decision, individuals have to process information faster and thus prefer the riskier choice (Chandler and Pronin 2012). The risk of pledge is higher than nonpledge because no matter the expected utility of nonpledge is positive or negative, the value should be a fixed number while the expected utility of pledge changes according to the progress as well as the outcome of the project.

Last but not least, time-based unavailability is a costly signal for creators. Creators take more risk to shorten the funding period because shorter funding period means fewer potential backers can notice the focal project and thus fewer backers can pledge. Only if creators are very confident about the quality of their project and believe their project can attract enough funding in a relatively short time, they dare to shorten the funding period. Otherwise, their project will fail, and creators cannot get any fund. Therefore, backers tend to believe the time-based unavailability can signal the real inherent quality. Based on the arguments above we hypothesize that,

H2: Time-based unavailability is positively associated with crowdfunding success.

2.4 Interacting Effect of Time-Based Unavailability and Quantity-Based Unavailability

Time-based unavailability can also alter how individuals interpret quantity-based unavailability. As mentioned before, single quantity-based unavailability has already been interpreted as sales tactics, which may not attract potential backers. However, if quantity-based unavailability is accompanied by the time-based unavailability, potential backers will change their interpretation. Time-based unavailability brings great risk for creators to raise capital because some crowdsourcing platforms such as Kickstarter use the all-or-nothing rule. If project creators do not raise enough funds to meet the funding goal, they will get nothing. Projects with higher time-based unavailability should have higher internal quality so when the quantity-based unavailability is also high, individuals will tend to believe the project is authentically scarce. Thus, we hypothesize that,

H3: The impact of time-based unavailability on crowdfunding success is stronger for projects with higher quantity-based unavailability.

3 Methods and Data

The platform we study in this research is Kickstarter, which maintains a global crowdfunding platform focused on creativity and merchandising. Kickstarter allows entrepreneurs to create projects on the web and attract backers to invest for the promised rewards listed in the main page of the focal project. For the project itself, entrepreneurs can write in plain text to describe their project and also upload pictures as well as videos to provide a more vivid blueprint. In the description part, entrepreneurs can design a reward system that explains how much investments correspond to what rewards and how long the backers can receive the rewards. Even though the description of a project is given in detail, it is still impossible for either the Kickstarter platform or the individual backers to quantify the quality of a project. Thus, potential backers tend to further rely the background information of the creator (the entrepreneurs) of the project. Creators have the autonomy to disclose their social network accounts, such as, Facebook, Twitter, and Instagram and other websites related to the project or themselves. Backers can, therefore, get more information to access both the capabilities and the motivation of the focal creators to realize the projects. Besides, Kickstarter also discloses the backing history and creating history of the focal creator. Backers can easily know how many projects the creator back and create on the platform. To provide a direct channel for creators and backers communicate, Kickstarter offer the comment section for each project webpage where backers can make comments and request more information about the project and the creator can decide which comment to respond and how to respond.

We crawl the information of more than 400,000 projects by more than 330,000 creators on the Kickstarter platform from 2009 to 2019. Our sample size is close to the official number of all projects on Kickstarter (Kickstarter, 2019). The missing projects are those whose project information is deleted by creators or Kickstarter platform.

Fig. 1. An example on quantity-based unavailability and time-based unavailability

Figure 1 gives an example on quantity-based unavailability and time-based unavailability[4]. We measure quantity-based unavailability as the number of reward categories with quantity limit and then divided by total number of reward categories and funding period is related to the time-based unavailability. Except for percentage variables and dummy variables, all other variables are log-transformed to get a more normalized distribution (Table 1).

Table 1. Explanation of variables

Variable names	Explanations
Dependent variables	
Success	1 if the project is fully funded and 0 otherwise
Independent variables	
Quantity-based unavailability	The percentage of rewards category with quantity constraints
Time-based unavailability	The length of funding period in days multiplied by −1 (a larger number represents higher unavailability)
Control variables	
Number of words	The number of words the creator uses to describe the project
Number of pictures	The number of pictures the creator uses to describe the project
Number of videos	The number of videos the creator uses to describe the project
Funding goal	The funding goal of the project (USD)
Staff pick	1 if the project is recommended by the Kickstarter staff and 0 otherwise
Number of total websites	The number of external websites disclosed by the creator
Number of collaborators	The number of collaborators in the project
Number of projects backed	The number of projects the creator has backed on Kickstarter
Number of projects created	The number of projects the creator has created on Kickstarter
Number of pre-launch total comments	The number of comments made by backers before the deadline of the fundraising period
Number of reward categories	The number of choices the creator provides for the backers to invest
Estimated days to deliver	The estimated days to deliver rewards from the deadline of the fundraising period

---

[4] https://www.kickstarter.com/projects/cavinbounce/lights-coma-action (accessed August 2019).

From Table 2, we can see that the success rate in our sample is 38%.

Table 2. Descriptive statistics

Variables	Mean	S.D.
Success	0.38	0.49
Time-based unavailability	−3.50	0.37
Quantity-based unavailability	0.39	0.37
Number of words	6.18	0.97
Number of pictures	1.90	0.83
Number of videos	0.52	0.37
Funding goal	8.63	1.68
Staff pick	0.060	0.24
Number of total websites	1.54	1.59
Number of collaborators	0.060	0.24
Number of projects backed	0.92	1.22
Number of pre-launch comments	0.74	1.27
Number of reward categories	2.00	0.62
Estimated days to deliver	4.03	1.92

4 Results

In the empirical testing part, we use the Logit regression model and include year fixed effect and project category fixed effect. The variable "Time-based # Quantity-based" represents the interacting effect. The results are shown in Table 3. Basically, all our three hypotheses are supported.

Table 3. Regression results

Variables	Success	Success
Quantity-based unavailability	−0.203***	0.602***
	(0.014)	(0.115)
Time-based unavailability	0.526***	0.446***
	(0.013)	(0.017)
Time-based # Quantity-based		0.233***
		(0.033)
Number of words	0.023***	0.023***

(continued)

Table 3. (*continued*)

Variables	Success	Success
	(0.007)	(0.007)
Number of pictures	1.157***	1.158***
	(0.010)	(0.010)
Number of videos	0.348***	0.348***
	(0.015)	(0.015)
Funding goal	−0.813***	−0.813***
	(0.004)	(0.004)
Staff pick	0.997***	0.997***
	(0.019)	(0.019)
Number of total websites	0.054***	0.054***
	(0.003)	(0.003)
Number of collaborators	0.689***	0.691***
	(0.022)	(0.022)
Number of projects backed	0.369***	0.369***
	(0.004)	(0.004)
Number of pre-launch total comments	0.961***	0.962***
	(0.005)	(0.006)
Number of reward categories	0.624***	0.625***
	(0.011)	(0.011)
Estimated days to deliver	−0.021***	−0.021***
	(0.003)	(0.003)
Constant	2.995***	2.707***
	(0.155)	(0.160)
Observations	408,380	408,380
Project country location fixed effect	YES	YES
Launched year fixed effect	YES	YES
Project category fixed effect	YES	YES
Pseudo R2	43.73	43.74

5 Limitations and Concluding Remarks

Our paper is subject to several limitations. First, the endogeneity hampers the validity of the casual inference in our regression analysis. Because both the quantity-based unavailability and time-based unavailability are determined by the project creator, it is very likely that these two factors are inherently correlated with the attributes of the project, for example, the quality of the project. Thus, the likelihood of success is affected. To address this concern, we are going to use multiple-step propensity score matching to keep the projects with different level of attributes comparable to each other. After matching, the heterogeneity stemming from the choice of project creator is mitigated. Second, measurement error may exist. For example, we measure the quantity-based unavailability as the percentage of rewards category with quantity constraints. Other measurements, such as, a dummy variable on whether there is any category with quantity constraints or, much more strictly, whether all the categories are with quantity constraints, may be considered. In the future work, we should consider different measurements for robustness check at different levels. Third, our sample still misses a portion of projects and there are some other rewarded-based crowdfunding platforms so we should try to collect more comprehensive samples to justify the generalizability. However, from the perspective of the Kickstarter platform, our sample is very close to the whole population. Besides, future research may incorporate experiment part to investigate more detailed underlying mechanisms. Our research facilitates further understanding of individuals' decision-making process on the online crowdfunding platform and at the same time identifies the boundary condition of unavailability to be effective. Our finding suggests that quantity-based unavailability (costless signal of quality) cannot attract potential backers but time-based unavailability (costly signal of quality) can. We also show that time-based unavailability can enhance the reliability of quantity-based signal by investigating the interacting terms.

References

Aggarwal, P., Vaidyanathan, R.: Use it or lose it: purchase acceleration effects of time-limited promotions. J. Consum. Behav. **2**(4), 393–403 (2003)

Agrawal, A., Catalini, C., Goldfarb, A.: Crowdfunding: geography, social networks, and the timing of investment decisions. J. Econ. Manage. Strategy **24**(2), 253–274 (2015)

Belleflamme, P., Lambert, T., Schwienbacher, A.: Crowdfunding: tapping the right crowd. J. Bus. Ventur. **29**(5), 585–609 (2014)

Brock, T.C.: Implications of commodity theory for value change. In: Psychological foundations of attitudes, pp. 243–275. Elsevier, (1968)

Burtch, G., Ghose, A., Wattal, S.: Cultural differences and geography as determinants of online pro-social lending. MIS Q. **38**(3), 773–794 (2014)

Cachon, G.P., Gallino, S., Olivares, M.: Does adding inventory increase sales? evidence of a scarcity effect in us automobile dealerships. Manage. Sci. **65**(4), 1469–1485 (2018)

Chaiken, S.: Heuristic versus systematic information processing and the use of source versus message cues in persuasion. J. Pers. Soc. Psychol. **39**(5), 752 (1980)

Chandler, J.J., Pronin, E.: Fast thought speed induces risk taking. Psychol. Sci. **23**(4), 370–374 (2012)

Cialdini, R.B.: Influence: The Psychology of Persuasion. HarperCollins, New York (2007)

Clee, M.A., Wicklund, R.A.: Consumer behavior and psychological reactance. J. Consum. Res. **6**(4), 389–405 (1980)

Colombo, M.G., Franzoni, C., Rossi-Lamastra, C.: Internal social capital and the attraction of early contributions in crowdfunding. Entrepreneurship Theor. Pract. **39**(1), 75–100 (2015)

Cui, R., Zhang, D.J., Bassamboo, A.: Learning from inventory availability information: Evidence from field experiments on amazon. Manage. Sci. **65**(3), 1216–1235 (2018)

Dai, H., Zhang, D.J.: Prosocial goal pursuit in crowdfunding: Evidence from kickstarter. J. Mark. Res. **56**(3), 498–517 (2019)

Fromkin, H.L., Snyder, C.R.: The search for uniqueness and valuation of scarcity. In: Social Exchange, pp. 57–75. Springer (1980)

Hong, Y., Hu, Y., Burtch, G.: Embeddedness, pro-sociality, and social influence: Evidence from online crowdfunding. MIS Q. **42**(4), 1211–1224 (2018)

Inman, J.J., McAlister, L.: Do coupon expiration dates affect consumer behavior? J. Mark. Res. **31**(3), 423–428 (1994)

Kahneman, D., Tversky, A.: Prospect theory: an analysis of decision under risk. Econometrica **47**(2), 263–291 (1979). https://doi.org/10.2307/1914185

Kremer, M., Debo, L.: Inferring quality from wait time. Manage. Sci. **62**(10), 3023–3038 (2015)

Li, G., Wang, J.: Threshold effects on backer motivations in reward-based crowdfunding. J. Manage. Inf. Syst. **36**(2), 546–573 (2019)

Lin, M., Prabhala, N.R., Viswanathan, S.: Judging borrowers by the company they keep: friendship networks and information asymmetry in online peer-to-peer lending. Manage. Sci. **59**(1), 17–35 (2013)

Lin, M., Viswanathan, S.: Home bias in online investments: an empirical study of an online crowdfunding market. Manage. Sci. **62**(5), 1393–1414 (2015)

Liu, D., Brass, D., Lu, Y., Chen, D.: Friendships in online peer-to-peer lending: pipes, prisms, and relational herding. MIS Q. **39**(3), 729–742 (2015)

Liu, Q., Van Ryzin, G.J.: Strategic capacity rationing to induce early purchases. Manage. Sci. **54**(6), 1115–1131 (2008)

Lukkarinen, A., Teich, J.E., Wallenius, H., Wallenius, J.: Success drivers of online equity crowdfunding campaigns. Decis. Support Syst. **87**, 26–38 (2016)

Lynn, M.: Scarcity effects on value: a quantitative review of the commodity theory literature. Psychol. Mark. **8**(1), 43–57 (1991)

Lynn, M.: Scarcity's enhancement of desirability: the role of naive economic theories. Basic Appl. Soc. Psychol. **13**(1), 67–78 (1992)

Mollick, E.: The dynamics of crowdfunding: an exploratory study. J. Bus. Ventur. **29**(1), 1–16 (2014)

Sharma, E., Alter, A.L.: Financial deprivation prompts consumers to seek scarce goods. J. Consum. Res. **39**(3), 545–560 (2012)

Stock, A., Balachander, S.: The making of a "hot product": a signaling explanation of marketers' scarcity strategy. Manage. Sci. **51**(8), 1181–1192 (2005)

Van Herpen, E., Pieters, R., Zeelenberg, M.: When demand accelerates demand: trailing the bandwagon. J. Consum. Psychol. **19**(3), 302–312 (2009)

Verhallen, T.M., Robben, H.S.: Scarcity and preference: an experiment on unavailability and product evaluation. J. Econ. Psychol. **15**(2), 315–331 (1994)

Worchel, S., Lee, J., Adewole, A.: Effects of supply and demand on ratings of object value. J. Pers. Soc. Psychol. **32**(5), 906 (1975)

Xu, J.J., Chau, M.: Cheap talk? The impact of lender-borrower communication on peer-to-peer lending outcomes. J. Manage. Inf. Syst. **35**(1), 53–85 (2018)

Young, D.L., Goodie, A.S., Hall, D.B., Wu, E.: Decision making under time pressure, modeled in a prospect theory framework. Organ. Behav. Hum. Decis. Process. **118**(2), 179–188 (2012)

Younkin, P., Kuppuswamy, V.: The colorblind crowd? Founder race and performance in crowdfunding. Manage. Sci. **64**(7), 3269–3287 (2017)

The Impact of Blockchain on Medical Tourism

Abderahman Rejeb[1] (ORCID), John G. Keogh[2] (ORCID), and Horst Treiblmaier[3](✉) (ORCID)

[1] Doctoral School of Regional Sciences and Business Administration, Széchenyi István University, Győr 9026, Hungary
[2] Henley Business School, University of Reading, Greenlands, Henley-on-Thames RG9 3AU, UK
[3] Department of International Management, Modul University Vienna, 1190 Vienna, Austria
Horst.Treiblmaier@modul.ac.at

Abstract. Medical tourism has witnessed significant growth over the last decade. This nascent sector creates a new tourist class with access to affordable healthcare services by combining healthcare services with tourism and hospitality. Information technology is an essential factor, which can enable the growth of medical tourism. Technology enables the search process for information about the available services, costs, hospitality, tourism and post-treatment options. However, these technologies are primarily legacy systems and lack interoperability. Several questions arise, including the potential patient-tourist ability to verify crucial factors such as the quality of care and the credentials of the medical professionals and medical facilities. Moreover, questions arise regarding patient-doctor trust, procedure and risk transparency, medical record privacy and other health-related hazards in specific procedures. In this conceptual paper, we investigate the potential benefits of Blockchain technology to address some of the open questions in medical tourism. We conclude that Blockchain technology can benefit medical tourism, and we lay the foundation for future research.

Keywords: Medical tourism · Blockchain technology · Trust & transparency · Privacy · Efficiency

1 Introduction

Over the past two decades, medical tourism has grown in prominence, appeal and acceptance as consumers seek faster and cheaper medical interventions. Drivers of this development include the high costs of private medical insurance, low reimbursement rates for specific procedures, lack of local expertise, and long wait times in public healthcare. The scope of medical tourism ranges from minor dental procedures to cosmetic surgery and significant interventions such as an organ transplant. Medical tourism positively contributes to economic growth in adjoining sectors, including tourism, transportation, pharmaceutical industry, and hospitality. The magnitude of medical tourism is reflected in the fast-growing number of medical tourists estimated to be 20–24 million cross-border patients worldwide [1]. This steadily growing niche market is intended to meet travelers' needs for quality care with affordable medical interventions and treatments, as

© Springer Nature Switzerland AG 2020
K. R. Lang et al. (Eds.): WeB 2019, LNBIP 403, pp. 29–40, 2020.
https://doi.org/10.1007/978-3-030-67781-7_4

well as the provision of highly specialized healthcare facilities. The number of countries which are developing their medical tourism markets has increased, and their services are expanding with competitive pricing [2]. Moreover, the globalization of medical tourism is aided by higher disposable incomes, technology transfer and growing competition in this lucrative segment. Additionally, the globalization of the 'sharing economy' lowers costs for local transportation (e.g., Uber) and provides an alternative, cost-effective accommodation options (e.g., Airbnb) [3].

The introduction of new technologies in travel and tourism introduces new consumer-centric tools. Research in the tourism field underlines the impact of technology and contributes to the continuous development of strategies to increase medical travel satisfaction. For example, information technology has been a critical factor in providing foreign medical tourists with easy access to information about the treatments and interventions that private hospitals in Thailand and India can offer [4]. Moreover, several information systems can be profitably integrated into the medical tourism industry, such as point-of-sale (POS) systems for cosmetic services [5], electronic health records systems (EHR) [6], and destination management organisations (DMOs) websites [7]. Blockchain has become a promising technology for driving large-scale societal and economic change. Blockchain is defined as a "digital, decentralized, and distributed ledger in which transactions are logged and added in chronological order with the goal of creating permanent and tamperproof records" [8]. The heightened popularity of blockchain stems from its ability to create a trustworthy network where value can be exchanged between peers. The technology has extensive, industry-spanning possibilities for applications [9, 10]. Blockchain technology has received increasing attention from academic researchers and industry practitioners who aim to solve persistent problems in many areas. Even though the impact of blockchain on tourism has been identified as an important research topic that can trigger essential transformations [11], the nascent area of medical tourism still lacks rigorous research. With the rapid digitization of healthcare, the introduction of blockchain technology is expected to have significant implications on healthcare delivery services for medical tourists. For example, medical tourism providers face the challenge of convincing potential customers of the quality of medical care centres and the safety of health outcomes [12]. In this conceptual paper, we discuss how blockchain technology can enable patient travelers to gain in-depth knowledge and enhanced trust in medical tourism destinations. In doing so, we lay the foundation for future theoretical and empirical research.

2   Definitions and Drivers of Medical Tourism

Goodrich and Goodrich [13] describe medical tourism as a plan to co-promote healthcare and tourism services. As such, the health-conscious consumer becomes both a tourist and a patient to obtain preventative healthcare assessments or healthcare treatments such as cosmetic, dental or even invasive surgical interventions. According to Carrera and Bridges [14], medical tourism involves systematic planning, maintaining and restoring one's physical and mental health condition. The search for medical services is, therefore, the primary motivator for medical tourists' travel decisions. More broadly, the final decisions made by aspirational medical tourists are influenced by their online search for trusted, safe, timely and cost-effective healthcare options and services. The primary drivers for searching for alternative healthcare and medical intervention options are affordability, accessibility and availability. In many countries, particularly those located in developed regions in the world, public healthcare often provides essential services while private medical insurance coverage carries a high cost and may penalize those with pre-existing medical conditions. Research from Canada shows that public healthcare often has extended wait times for interventions that can significantly impact patients' health and wellbeing [15]. Research on women's attitudes toward cosmetic surgery in Australia reveals that peer pressure, media exposure and global views on appearance ideals are drivers for undergoing such treatment while social acceptance of cosmetic surgery functions as an enabler [16].

Social pressure and universal ideals of beauty can create an obsession with physical appearance. In turn, this leads to elective, minimally invasive procedures (e.g., remorseful tattoo removal, microdermabrasion) or more invasive cosmetic procedures on the face and neck (e.g., rhinoplasty, otoplasty, blepharoplasty, mentoplasty, rhytidectomy) and body (e.g., abdominoplasty, brachioplasty, mammoplasty) and specific male and female genitalia enhancements [16]. In this 'vanity' category of elective cosmetic surgery, the medical tourist may prefer to be away from their social and work networks during the recovery, which may include wearing bandages for several weeks or include significant swelling and discomfort. Additionally, medical tourism may include patients with an 'acquired defect' resulting from various types of trauma (e.g., accident deformity or burn) or from cancer. Furthermore, patients may have a congenital disability, such as a cleft lip, cleft palate or other physical deformities. Finally, another market segment includes gender reassignment or correction surgeries.

Affordability

Medical tourists must consider the affordability of international travel, how favorable the currency exchange rates are and the increasing sophistication of medical care in some developing countries [17]. Since the technological gap has been narrowed globally, the affordability of medical care has become a decisive factor for shifting the patterns and flows of medical travellers. Additionally, several medical destinations are responsive to the evolving needs of the 'tourist-patient' for affordable medical treatments, allowing them to gain a cost advantage and a competitive position in the global market. As a case in point, spinal surgery could cost a patient about $70,000 in the United States. However, the same medical intervention, coupled with a five-day stay in a private recovery room, costs $4,700 in a recognized hospital in Thailand [18]. The significantly lower costs apply to a growing array of healthcare services provided throughout Central Europe and Asia [19].

Accessibility and Availability

Two important drivers of medical tourism are the accessibility and availability (including wait times) of specific healthcare services in patients' home countries. More specifically, the undertaking of medical travel is either due to the absence of the required medical services in the patients' home country or to their non-availability at a certain point in time and need. Furthermore, medical tourism might be an option when treatment procedures are delayed due to long waiting times, priority listings, and the paucity of organs necessary for transplant operations. Dawn and Pal [20] highlight that most medical tourists from industrialized regions such as the United Kingdom, Japan, the United States, and Canada receive their treatments abroad because of the long waiting times for medical consultations and interventions. Similarly, the Japanese healthcare system's capacity to handle the medical demands of the country's ever-growing elderly population drives medical tourism. For instance, many Japanese firms send their employees to neighbouring countries for yearly medical checkups. Medical tourism, such as this, showcases the growing importance of organ transplant tourism, which saves thousands of lives every year.

3   The Potentials of Blockchain Technology in Medical Tourism

Blockchain technology has numerous characteristics such as immutability, trust, transparency and enhanced security, which can significantly impact business processes as well as whole industries [21]. In this paper, we primarily focus on those characteristics which can impact the medical tourism industry. In Fig. 1, we summarize four essential areas of focus, namely, *disintermediation, trust and transparency*, *digitization and interoperability* and *privacy*. We will elaborate on these in the sections below.

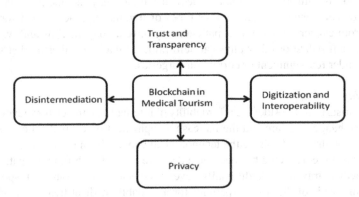

Fig. 1. The impact of blockchain on medical tourism

3.1   Enabling Disintermediation

Increasing demand for medical tourism has paved the way for healthcare intermediaries to become facilitators between international patients and medical service providers [12]. Intermediaries provide a wide array of value-added services for this new class of tourist-patient. The emergence of intermediaries is mainly due to medical tourists' lack of technical knowledge and their inability to assess the quality, suitability, and benefits of a medical-tourism destination [22]. As such, potential patients experience significant information asymmetry about the specific healthcare facilities and rely on intermediaries to answer queries and arrange their travel and hospitality. The activities carried out by intermediaries include the matching of patients with appropriate medical care providers. It may extend to arranging specific travel procedures (e.g., specialized equipment, prescription drugs permits, or nurse-assisted travel), as well as the arranging visas, flights, accommodations, treatments, and post-operative care during recuperation [23].

Traditional travel agencies maintain a dominant position in the medical tourism industry, which can lead to opportunistic behavior. The role of medical tourism intermediaries has led to a paternalistic model of relationships, rather than the establishment of close connections between medical institutions and their patients [24]. Medical tourism packages offered by these entities are often delivered at inflated prices, which are marked up from the actual costs of the healthcare service providers. Moreover, medical intermediaries are very likely to restrict available options due to their affiliation with specific medical providers to whom they exclusively send their patients [25]. Being specialized in a particular target market or service, intermediaries might refer patients to inappropriate medical providers with additional mark-ups and high referral fees. More troubling, these medical tourism intermediaries are unlikely to be held legally accountable for any failures, since they are not healthcare providers themselves [26]. Blockchain technology can significantly lessen the asymmetric power of medical tourism intermediaries. For example, blockchain technology can be embedded in applications that support medical tourists to engage in direct, interactive communication with foreign healthcare service providers. In doing so, patients can objectively verify their credentials such as specific surgical qualifications, practitioner certifications and procedural costs. This can lead to the elimination of unnecessary costs and reduce patient dependence on exclusive arrangements through intermediaries to arrange all aspects of their medical travel and tourism. Additionally, streamlining the medical tourism infrastructure by removing or lessening the power of non-value adding intermediaries will benefit the sector [27]. The technology, therefore, holds an untapped potential to unlock new value, since it reduces information asymmetry by enhancing information transparency and knowledge about medical facilities. The use of blockchain technology can address several issues by enabling a fully disintermediated approach or enhancing aspects of the intermediation where medical travel agents are incentivized to specialize and offer superior value-added services to their potential customers [28].

3.2 Enhancing Trust and Transparency

Sirdeshmuhk, Singh, and Berry [29] define trust as the expectation held by the consumer that the service provider is dependable and can be relied upon to deliver its promises. In the context of medical tourism, this implies that the (mostly foreign) health service provider is reliable, that their medical professionals are competent, and that the entire medical institution is adept at improving the health outcomes of their patients. The proliferation of informative medical tourism websites creates greater awareness of healthcare services. Rather than consulting with a medical tourism agency or contacting a healthcare facility directly, medical tourists favour online methods to quickly obtain and assess healthcare services and prices [30]. The use of ICTs supports the patient-tourist in their decision-making process and may assist them in negotiating medical packages. Medical tourism businesses strive to attract customers through new marketing channels, media coverage and social media platforms such as YouTube, Facebook and Twitter.

ICTs are essential enablers of medical tourism, especially in the initial phase of information search and medical facility selection. According to Connell [12], the biggest hurdle that medical tourism providers have to face is the challenge of persuading potential patients about their qualifications, quality of pre, and post-procedure services, and overall patient safety. Medical facilities signal their qualifications through various online and offline channels to reduce information asymmetry between them and the potential patient. Potential patients are signal receivers but do not have the expertise or necessary tools to verify these 'unobservable' qualifications from the medical facility [31]. If qualifications cannot be verified, a potential patient lacks trust in the medical providers' signals of their unobservable features such as quality of care and skill levels of specialists. Blockchain technology can be used to link health facilities qualification claims to their various certification bodies and authorities. In this regard, blockchain technology facilitates the verification process of credentials, including certifications and qualifications of the facilities, doctors or specialists, and the authentication of online reviews. The latter is especially useful because potential patients often feel uncertain about post-procedural outcomes or unfulfilled expectations.

The reliability of online health information is often questionable, and individuals are cautious when seeking medical information online [24]. Moreover, most medical websites aim to influence consumers' cognitive, affective, and behavioural dimensions [32] while ignoring their real needs. Some websites may entice customers to purchase medical travel packages, which may include a high-risk procedure carried out in non-accredited health institutions by untrained professionals [33].

With all of these challenges in mind, the medical tourism industry is primed to become a principal beneficiary of blockchain technology. In line with Pilkington [34], the use of blockchain in medical tourism can enhance patient trust through improved information transparency. Using blockchain technology in consumer-facing applications will allow potential medical tourists to make reasonable and well-informed decisions. In the planning stage, prospective medical tourists can use blockchain-enabled systems to optimize their search for medical tourism services by receiving detailed, authentic, and verified information about medical tourism institutions. This includes verified curriculum vitae of healthcare staff, institutions' accreditations, and service or procedure certifications. The benefits of this approach are twofold: first, travellers' confidence will be strengthened before entering the unknown medical system. Second, medical tourism providers can build brand equity for their customers from the beginning and sustain a consistent brand image [35]. The integration of blockchains in the medical tourism industry can provide credible and in-depth medical and non-medical information. Blockchain-enabled systems can help medical tourists make a more informed decision, regardless of their socio-economic, cultural, or linguistic backgrounds [36].

3.3 Enhancing Digitization and Interoperability

The operational mechanism of health-related systems has several significant limitations. EHRs are still scattered over different healthcare systems and 'siloed' in nature [37]. This results in inaccessibility problems and inconsistency of medical information. Similarly, the lack of integration of EHR management systems [38] binds medical patients with specific foreign health service providers. This implies that if medical tourists want to terminate their involvement with a particular foreign healthcare provider, then they would not be able to transfer their medical information to their new provider, as they lack access to the records system. As a consequence, interoperability, security, and privacy issues resulting from the online exchange of health-related records should not be overlooked as being part of the medical tourism experience.

By empowering a patient-centric model of information handling, blockchain technology can provide a holistic and comprehensive assessment of an individual's health. This can be achieved through the linking and active tracing of patients' entire medical history, sophisticated record management of health documents, and increased control of data access. A medical tourism ecosystem based on blockchain technology allows overseas patients to maintain an increased level of control over their health data and to take an active role in ensuring communication between their overseas healthcare providers and their local physicians [39]. As a result, blockchain can address important interoperability issues inherited from existing health information technology systems, enabling medical tourists, foreign healthcare providers, and other stakeholders to have a borderless, timely, and secure exchange of health data [40]. Furthermore, blockchain can significantly improve the quality of healthcare service delivery to medical tourists. This can be achieved by using cryptocurrencies and medical tourists can settle payments quickly and securely while minimizing transaction costs [11].

As medical tourists return home, delays and discontinuities in the patients' continuum of care may occur due to a lack of access and interoperability to their overseas health records. Moreover, there is a higher risk of errors in diagnoses if the patient's full medical records are not available. Furthermore, recommended treatments may include drugs that are not available in their home countries. Blockchain technology can create an effective solution to the enduring problem of fragmentation in medical records and the alignment of prescribed medications to those which are approved and readily available in their home countries. Importantly, blockchain architecture also guarantees that patients have secure access and control over their encrypted health records. Importantly, blockchains immutability feature ensures that patients are unable to change, remove, or add any health-related information to these records. Therefore, patients' medical records on a blockchain are secured, encrypted and have authentication mechanisms that preserve the integrity of medical information. These capabilities have already caught the attention of policymakers who consider technology as an enabler for health tourism [34].

3.4 Alleviating Privacy Concerns

One integral component of medical tourism is patient privacy. Medical tourism patients often entrust their sensitive records to untrusted healthcare providers or intermediaries [12]. A survey among American medical tourists showed that 'privacy and confidentiality of treatment' was the second most crucial element they considered, behind treatment costs [51]. In an Asia-Pacific study on wellness tourism, privacy was identified as a crucial factor for higher-paying guests [42]. As far as their medical records sensitivity is concerned, some tourist-patients feel more confident when they know that their privacy will be protected, regardless of the geographic remoteness of their medical tourism destination. However, this 'remoteness' might only appease patients' privacy concerns because confidentiality problems still exist that can run counter to the expectations of medical tourists. Besides, some prospective medical tourists might have doubts about a foreign country's readiness to enforce appropriate privacy policies and protect private medical information. This distrust is aggravated by the use of modern ICTs in medical tourism. Previous research by Laric and Pitta [43] suggests that technological advances and digitization of medical information intensify, rather than alleviate, patients' privacy concerns. While these technologies have significantly facilitated the recall of information on medical procedures and details, they have also opened up a new arena for privacy intrusions. Wagle [44] provided evidence that websites regularly display patient information in the form of testimonials and photographs, along with the type of treatment administered. Even though this information serves as a marketing tool for foreign healthcare service providers, there is a strong likelihood that their use is without the tourist patients' knowledge and consent [45].

Destinations which promote surrogacy—a medical tourism subset referring to the gestation of a baby by one mother for another [46]—have implemented practices that exacerbate privacy issues. For example, the US state of Georgia facilitates surrogacy procedures by exhibiting a database of surrogate mothers with photographs, a practice that typifies a very pronounced act of privacy violation in many countries [26]. Moreover, according to Angst and Agarwal [47], the tension between technology and confidentiality amplifies some patients' reluctance to use electronic health record systems (EHR). The possible reasons for this are that patients fear that their data will be divulged, leaked, or stolen and that they could fall victim to identity theft [48]. Therefore, the privacy risks arising from existing computerized hospital systems, medical databases, and EHRs pave the way for privacy violations in medical tourism.

Blockchain technology can be viewed as a 'privacy-by-design' solution for the many privacy issues resulting from the digitization of medical information [49]. As such, privacy and data protection mechanisms are embedded in the blockchain system from the inception of the system's design, rather than being an add-on feature. The confluence of decentralization and distributedness embodied in blockchain technology allows the patient-tourist to have more control over their personal medical information. With control and ownership, the medical traveller's privacy concerns tend to ease because they have full knowledge of the flow and use of their private information [50]. Also, foreign health entities will be able to communicate privacy-enhancing features and thus show a strong commitment to protecting the confidentiality of overseas patient-tourists, which, in turn, makes patients feel more confident, relaxed, and even potentially willing to disclose their private information. It should be noted that the potential information transparency of blockchain technology, enabled by shared data access, does not necessarily entail the goal of privacy. Cryptographic identity schemes offer secure confidentiality through anonymity or pseudonymity and the unlinkability of transactions. The built-in privacy of blockchain allows patients to be self-sovereign over sensitive information, which could be shared partly or wholly, temporarily or permanently, and restrictedly or unrestrictedly [51]. For instance, the use of private blockchains in medical tourism entails that the patient-tourist information and data will be controlled by a permissioned mechanism that assigns different access rights to foreign medical entities. The privacy of medical tourists will no longer be an issue even after receiving a medical service because their control is long-lasting.

4   Conclusion and Future Research

The desire to prolong the quality of life, restore health and find an enjoyable leisure experience boosts the global development of medical tourism. This nascent industry segment fostering international travel by patients seeking medical treatment is booming. Medical tourists' needs range from non-invasive rejuvenation to risky and invasive interventions for serious medical conditions that often require complex surgeries [26]. This multi-disciplinary service segment is triggered by push factors that urge people to engage in a medical travel experience. This has led to the possibility of accessing highly regarded health institutions, medical professionals, sophisticated technologies and quality treatments. Moreover, these services and procedures are often viewed as being more affordable, accessible and available than the comparable treatments in the patients' home nation. Medical tourism is also appealing for those seeking to accommodate their recuperation in comfortable physical surroundings located in luxurious and distant therapeutic venues, as well as to participate in entertaining activities (e.g. sightseeing, food, cultural visits) during their stay. Despite the vital role of medical tourism in responding to many patients' critical and sensitive needs, uncertainty still permeates every part of the medical travel process. At the planning stage, tourist patients often consult medical travel intermediaries who offer aid in preparing and arranging patients' itineraries. Intermediaries often facilitate the linking of medical tourism destinations to their prospective clients. Such intermediaries are not necessarily subject to external evaluation or accreditation, and many employ brokers who will connect medical tourists with the international hospital networks that they seek [12]. As a result, there are some unreliable, poor-quality medical products and services being marketed via the Internet (e.g., ill-considered and harmful cosmetic surgeries, ineffective treatments and ethically questionable organ transplantations). Furthermore, existing ICT systems in healthcare lack jurisdictional interoperability. As a result, there is a need to foster the growth of a more transparent and trusted health tourism segment. Healthcare providers and institutions need to re-engineer their care processes and reap the full benefits of health information technologies [52]. Importantly, privacy and security issues resulting from the digitization of medical records should not be overlooked during medical travel.

In this paper, we have responded to the significant challenges facing medical tourism by comprehensively elaborating on the possibilities of blockchain technology. Blockchain can solve several entrenched problems within the field of medical tourism. Disintermediation enabled by blockchain technology can fundamentally reshuffle power relationships among key players of the medical tourism industry by elevating the sense of autonomy experienced by tourist patients and lessening the reliance of patients on powerful intermediaries. Moreover, the 'trust-by-design' and transparency-enhancing features of blockchain technology allow prospective tourist patients to make informed decisions in the selection of their medical destinations. In facing the fragmentation and inefficiencies of health tourism systems, blockchain is a workable solution for improving and securing the flow of medical information and data between foreign health service providers and travelling patients. Information inconsistencies and interruptions in the healthcare continuum resulting from tourist patient's mobility can be avoided since blockchain reflects a permanent availability of their procedural interventions, transactions and overall medical history. Tourist patients can also benefit from blockchain's 'privacy-by-design' architecture because of its ability to bestow a solid sense of discretion and confidence that their personal information will be protected. This conceptual paper contributes to the current body of knowledge regarding the possibilities of blockchain technology in the medical tourism industry. It builds a better understanding of how emerging technological developments can be a viable solution for a number of medical tourism problems. Further research is needed to investigate the impact of blockchain technology rigourously, and to create a solid theory-based foundation. Blockchain technology is under constant development, and its potential to integrate with other technological advancements such as big data analytics, artificial intelligence (AI), machine learning (ML) and the Internet of Things (IoT) is vast but insufficiently understood. Further academic research is therefore called upon to contribute to the understanding of the positive and negative implications of blockchain and related technologies in medical tourism.

Acknowledgement. The publication of this work was supported by: EFOP- 3.6.1-16-2016-00017 (Internationalisation, initiatives to establish a new source of researchers and graduates and development of knowledge and technological transfer as instruments of intelligent specialisations at Széchenyi István University).

References

1. Patients Beyond Borders. Medical Tourism Statistics & Facts (2019)
2. Behrmann, J., Smith, E.: Top 7 issues in medical tourism: challenges, knowledge gaps, and future directions for research and policy development. Glob. J. Health Sci. **2**, 80 (2010)
3. Tsai, H., Song, H., Wong, K.K.F.: Tourism and hotel competitiveness research. J. Travel Tourism Mark. **26**, 522–546 (2009)
4. Levebvre, B., Bochatan, A.: The rebirth of the hospital: Heterotopia and medical tourism in Asia. Asia Tour, 113–124. Routledge (2002)
5. Cosmetisuite. Practice Management/ EHR Software For Plastic, Cosmetic, Hand Surgeons, Aesthetic, Dermatology & Medical Spa (2019)
6. Rezaei-Hachesu, P., Safdari, R., Ghazisaeedi, M., Samad-Soltani, T.: The applications of health informatics in medical tourism industry of Iran. Iranian J. Public Health **46**, 1147 (2017)
7. Qi, S., Law, R., Buhalis, D.: Usability of Chinese destination management organization websites. J. Travel Tourism Mark. **25**, 182–198 (2008)
8. Treiblmaier, H.: The impact of the blockchain on the supply chain: A theory-based research framework and a call for action. Supply Chain Manage. Int. J. **23**, 545–559 (2018). https://doi.org/10.1108/SCM-01-2018-0029
9. Wamba, F., Samuel, K.K., Robert, J., Bawack, R., Keogh, J.G.: Bitcoin, blockchain, and fintech: a systematic review and case studies in the supply chain. Product. Plan. Control **31**, 115–142 (2020)
10. Rejeb, A., Sűle, E., Keogh, J.G.: Exploring new technologies in procurement. Trans. Logistics: Int. J. **18**, 76–86 (2018)
11. Önder, I., Treiblmaier, H.: Blockchain and tourism: three research propositions. Ann. Tourism Res. **72**, 180–182 (2018)
12. Connell, J.: Medical tourism: Sea, sun, sand and surgery. Tourism Manage. **27**, 1093–1100 (2006)

13. Goodrich, G., Goodrich, J.: Healthcare tourism- an exploration study. Tourism Manage. **8**(3), 217–222 (1987)
14. Carrera, P.M., Bridges, J.F.P.: Globalization and healthcare: understanding health and medical tourism. Expert Rev. Pharmacoeconomics Outcomes Res. **6**, 447–454 (2006)
15. Patrick, J., Puterman, M.L.: Reducing wait times through operations research: optimizing the use of surge capacity. Healthcare Policy **3**, 75–88 (2008)
16. Sharp, G., Tiggemann, M., Mattiske, J.: The role of media and peer influences in Australian women's attitudes towards cosmetic surgery. Body Image **11**, 482–487 (2014). https://doi. org/10.1016/j.bodyim.2014.07.009
17. Swain, D., Sahu, S.: Opportunities and challenges of health tourism in India. In: Conference on Tourism in India–Challenges Ahead, 15:17. Citeseer (2008)
18. Smith, A.K.: Health care bargains abroad. Kiplinger's Personal Finance **66**, 65–68 (2012)
19. Bostan, I., Teodora, R., Cristina, L., Adriana, M. The current trends and opportunities in the industry of medical tourism. Revista de turism-studii si cercetari in turism: 58–63 (2016)
20. Dawn, S.K., Pal, S.: Medical tourism in India: issues, opportunities and designing strategies for growth and development. Int. J. Multi. Res. **1**, 7–10 (2011)
21. Treiblmaier, Horst: Toward more rigorous blockchain research: recommendations for writing blockchain case studies. Front. Blockchain **2**, 1–15 (2019). https://doi.org/10.3389/fbloc. 2019.00003
22. Legido-Quigley, H., McKee, M., Walshe, K., Suñol, R., Nolte, E., Klazinga, N.: How can quality of health care be safeguarded across the European Union? BMJ **336**, 920–923 (2008)
23. Lunt, N., Carrera, P.: Medical tourism: assessing the evidence on treatment abroad. Maturitas **66**, 27–32 (2010)
24. Lunt, N., Hardey, M., Mannion, R.: Nip, tuck and click: medical tourism and the emergence of web-based health information. Open Med. Inform. J. **4**, 1–11 (2010)
25. Herrick, D.M.: Medical tourism: global competition in health care. Natl. Center Policy Anal. **1**, 2–3 (2007)
26. Connell, John: Medical Tourism. Cabi, Wallingford (2011)
27. Raman, R.K., Varshney, L.R.:. Dynamic distributed storage for scaling blockchains, 1–19 (2007). arXiv preprint arXiv:1711.07617
28. Ehrbeck, T., Guevara, C., Paul D.: Mango Mapping the market for medical travel. McKinsey Q. 11 (2008)
29. Sirdeshmuhk, D., Singh, J., Berry, S.: Customer trust, value, and loyalty in relational exchange. J. Mark. **66**, 15–37 (2002)
30. Greenspan, R.: Internet High On Travel Destinations (2004)
31. Connelly, B.L., Certo, S.T., Ireland, R.D., Reutzel, C.R.: Signaling theory: a review and assessment. J. Manag. SAGE Publications Inc, 37(1), 39–67. https://doi.org/10.1177/014920 6310388419
32. Manaf, N.H., Abd, H.H., Kassim, P.N.J., Alavi, R., Dahari, Z.: Issues and challenges in medical tourism: an interdisciplinary perspective. Emerg. Mega-trends Asian Mark. **7**, 230–242 (2013)
33. Lee, H.K., Fernando, Y.: The antecedents and outcomes of the medical tourism supply chain. Tourism Manage. **46**, 148–157 (2015)
34. Pilkington, M.: Can blockchain technology help promote new tourism destinations? The example of medical tourism in Moldova. SSRN Electron. J. 1–8 (2017)
35. DeMicco, F.J.: Medical Tourism and Wellness: Hospitality Bridging Healthcare (H2H). CRC Press, Boca Raton, Florida (2017)
36. Lee, C.: Just what the doctor ordered: medical tourism. Monash Bus Rev **3**, 10–12 (2007)
37. Galen, D., et al.: Blockchain for Social Impact. Report, Stanford Graduate School of Business (2018)

38. Wang, C.J., Huang, A.T.: Integrating technology into health care: what will it take? JAMA **307**, 569–570 (2012)
39. Alleman, B.W., Luger, T., Reisinger, H.S., Martin, R., Horowitz, M.D., Cram, P.: Medical tourism services available to residents of the United States. J. Gen. Intern. Med. **26**, 492–497 (2011)
40. Linn, L.A., Koo, M.B.: Blockchain for health data and its potential use in health it and health care related research. ONC/NIST Use of Blockchain for Healthcare and Research Workshop, pp. 1–10. Gaithersburg, Maryland, United States, ONC/NIST (2016)
41. Singh, N.: Exploring the factors influencing the travel motivations of US medical tourists. Current Issues Tourism **16**, 436–454 (2013)
42. Kucukusta, D., Guillet, B.D.: Measuring spa-goers' preferences: a conjoint analysis approach. Int. J. Hosp. Manage. **41**, 115–124 (2014)
43. Laric, M.V., Pitta, D.A.: Preserving patient privacy in the quest for health care economies. J. Consum. Mark. **26**, 477–486 (2009)
44. Wagle, S.: Web-based medical facilitators in medical tourism: the third party in decision-making. Indian J. Med. Ethics **10**, 28–33 (2013)
45. Culnan, M.J., Armstrong, P.K.: Information privacy concerns, procedural fairness, and impersonal trust: an empirical investigation. Organ. Sci. **10**, 104–115 (1999)
46. Panitch, V.: Surrogate tourism and reproductive rights. Hypatia **28**, 274–289 (2013)
47. Angst, C.M., Agarwal, R.: Adoption of electronic health records in the presence of privacy concerns: the elaboration likelihood model and individual persuasion. MIS Q. **33**, 339–370 (2009)
48. Brown, C.L.: Health-care data protection and biometric authentication policies: comparative culture and technology acceptance in china and in the United States. Rev. Policy Res. **29**, 141–159 (2012)
49. Benchoufi, M., Ravaud, P.: Blockchain technology for improving clinical research quality. Trials **18**, 335 (2017)
50. Milne, G.R.: Privacy and ethical issues in database/interactive marketing and public policy: a research framework and overview of the special issue. J. Public Policy Mark. **19**, 1–6 (2000)
51. Lenz, R.: Managing Distributed Ledgers: Blockchain and Beyond (2019). Available at SSRN 3360655
52. Kellermann, A.L., Jones, S.S.: What it will take to achieve the as-yet-unfulfilled promises of health information technology. Health Aff. **32**, 63–68 (2013)

Business Analytics

Creating a Data Factory for Data Products

Chris Schlueter Langdon[1]([✉]) [iD] and Riyaz Sikora[2]

[1] Peter Drucker School of Management, Claremont Graduate University, 1021 N Dartmouth Ave Claremont, Claremont, CA 91711, USA
chris.langdon@cgu.edu
[2] Information Systems, College of Business, University of Texas at Arlington, P.O. Box 19437, Arlington, TX 76019, USA
rsikora@uta.edu

Abstract. Data is seen as the next big business opportunity. From a demand side, the popularity of artificial intelligence (AI) is growing and particularly deep learning requires large amounts of data. From a supply side, new technology, such as Internet of Things (IoT) sensors and 5G mobile communications, will greatly expand data generation. However, data has remained a challenge. In data analytics companies are struggling with too much time spent on data preparation. As of today, data preparation for analytics has largely remained handmade and made-to-order like cars before Henry Ford industrialized the auto business through productization of cars and parts, and factory automation. Similarly, for data analytics to become a bigger business, data has to be productized. First "data factories" are emerging to create such data products economically. This article introduces a framework to guide construction of a data factory: What are the key modules, why are they important, how is best practice evolving? The article is building on (a) a foundation and in-depth case studies in the literature, (b) current meta research and systematic literature reviews (SLRs), and (c) our own observations building a data factory. This real-world application uncovered the important additional steps of data rights management and data governance that may be less obvious from a computer science perspective but critically important from a business and information systems view.

Keywords: Data product · Data factory · Data sovereignty · Data governance · Data quality

1 Data as the Next Big Business

Data is promised to be the next big business (e.g., Wall 2019, Gartner 2018a). Investment banks, analysts and consultants further feed the frenzy with big revenue forecasts. In terms of data monetization opportunities, consultants McKinsey & Company estimate that car-generated data alone will be worth between US$450 billion and US$750 billion by 2030, less than two vehicle generations away (McKinsey 2016). Consumer data is already a business today. Google and Facebook live off the data that users create on their platforms. Almost all their revenue is from advertising, selling "eyeballs" and user engagement to advertisers.

© Springer Nature Switzerland AG 2020
K. R. Lang et al. (Eds.): WeB 2019, LNBIP 403, pp. 43–55, 2020.
https://doi.org/10.1007/978-3-030-67781-7_5

Lesser known consumer data companies are market researchers, like GfK, Ipsos and Nielsen - the top marketing research companies according to the (American Marketing Association 2018). Yet, the list of data vendors is far longer. A new 2019 Vermont law requires data brokers to be registered (Vermont 2018), and already a list of more than 120 companies has emerged (Melendez 2019).

All of the above is just the beginning. A big data boost is expected from IoT data (Internet of Things): IoT is essentially turning objects into websites. Historically, the Web and website tracking created a first wave of Big Data (which in turn created new technology to store and process it, such as Hadoop). Now ordinary objects are turned in to websites. For example, cars: connected and autonomous vehicles are projected to generate four terabytes (TB) of data a day (Krzanich 2016). Furthermore, this IoT boom is fueled by a confluence of trends in information systems, such as miniaturization of sensors like lidar (light detection and ranging sensor for autonomous cars), device technology, e.g., edge computing, and a new 5G cellular mobile communications standard.

1.1 The Problem: Data Isn't Scaling

A key mechanism to release value from data is analytics. With Websites it took tools like Google Analytics (Urchin) to benefit from Website tracking and attract advertising budgets. Google Analytics is mostly descriptive analytics. Far more value is generated from consecutive stages of predictive and prescriptive analytics (McKinsey 2018, Gartner 2018b). Examples include product recommendations using machine learning as an amplifier of word-of-mouth marketing (e.g., on Amazon and Netflix; see also Stern et al. 2009); and the application of deep learning or neural network methods across many domains for the recognition of text (sentiment analysis), picture (automatic license plate recognition, ALPR), video (autonomous vehicles) and speech recognition (Amazon's Alexa virtual assistant). Yet despite the media hype a quick review of time spent in data analytics projects reveals a big problem. Today, according to the literature more than 80% of the time budget of a data analytics project is spent on data wrangling - not with algorithms (Press 2016, Vollenweider 2016). Companies have gone from databases to data warehouses and now to data lakes (Porter and Heppelmann 2015) - and they seem to be drowning in it. Our own survey of data experts confirms the problem. If an analytics project is broken into the three phases of (a) data processing, (b) analytics modeling & evaluation, and (c) deployment, then timeshares are reported as 48%, 32% and 20% respectively (n = 65, our survey has been conducted in 2018 using a convenience sample of data experts at data science events for business – not academic conferences).

1.2 The Solution: Data Productization

These numbers confirm that data processing for AI remains handmade just like cars before Henry Ford industrialized auto making. Gottlieb Daimler invented the motor car in 1886, but it was Henry Ford who invented the modern auto business about 20 years later (Womak et al. 1990). He evolved auto making from a hand-made affair to mass production through automation, which made autos affordable for a big market. The moving assembly line is probably the most visible and striking feature. However, less obvious, for automation to work Ford critically required interchangeability of parts, which in turn required metrics (Clark and Fujimoto 1991). Parts had to be made to precise measurements so that all copies of a part were more or less similar in order to be attached to cars coming down the line quickly without lengthy calibration and refitting work. Mechanical engineering introduced the notion of tolerance as "the range of variation permitted in maintaining a specified dimension in machining a piece" (Webster 2019). Parts were specified ("specced") in engineering drawings or "blueprints" and then manufactured within precise tolerances to make them interchangeable.

So far, data has eluded proper measurements and is in need of productization (Crosby and Schlueter Langdon 2019, Glassberg Sands 2018). Data attributes have remained qualitative and subjective. Examples include fundamental properties, such as measures of size and quality. How to size data? What is big data? Is size measured in (i) bytes or (ii) population size or (iii) length of a time series - or all of the above? Our survey affirms the complication. All three dimensions seem to matter (bytes: 24%, population size: 31%, length of time series: 45%, n = 67). Same with quality: Without metrics data remains ambiguous like parts that may or may not fit, which will also inhibit data sharing and exchange. Akerlof has demonstrated how the lack of transparency of attributes or "asymmetry of information" between buyers and sellers will lower product quality (lemons) or prevent market exchange outright (Akerlof 1970).

2 Productization in "Data Factories"

In 2006 Clive Humby, a mathematician and architect of UK retailer Tesco's club and loyalty card, spoke of "data as the new oil" at the Association of National Advertisers' marketer's summit at Chicago's Kellogg School of Management (ANA 2006). The one part of his oil analogy, that data may be as valuable as oil has caught on - although data is not even used up in consumption like oil. The other part about the refining effort has not. Humby's analogy suggests that in order to prepare raw data into a refined data product for analytics applications - AI-ready data - it may take extensive refining and at an industrial scale with large platforms comparable to massive refineries for oil.

IT industrialization is certainly not a new phenomenon (Walter et al. 2007). And for plain data storage and processing this refinery analogy appears to correspond very well with observations in the field, specifically the explosive growth of the cloud business.

Launched with Amazon's Elastic Compute Cloud in 2006 the global public cloud service revenue for 2019 has been estimated to exceed US\$200 billion (Gartner 2018b). Furthermore, supporting Humby's scale argument, the business is already highly concentrated at an early age with only three hyperscalers dominating most of the business: Amazon's Web Services (AWS), Microsoft's Azure, and Google's Cloud Platform (GCP). As of end of 2018 these top 3 vendors accounted for 60% of the business, the top 10 for nearly 75% (Miller 2019).

2.1 Toward a Data Factory Framework

The conceptualization of our data factory framework builds on an established foundation. It has evolved in a multi-step investigation from (a) in-depth case study analysis in the literature and (b) systematic literature reviews (SLRs) to (c) our own observations building a data factory in practice. Figure 1 summarizes developments in the literature as the foundation of our refinements.

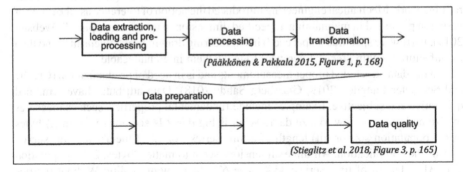

Fig. 1. Evolution of framework foundation in the literature

Pääkkönen & Pakkala present a first analysis of internal "data factories" using in-depth case studies of big data pioneers (2015). The authors dissect data operations at pioneers like Facebook and Netflix and establish that data preparation at these companies is a 'process' as "a series of actions or steps" (Webster) analogous to a 'factory' as "a set of [...] facilities for [...] making wares [...] by machinery" (Webster). This stepwise decomposition conforms with the evolution of information system capabilities toward modularization and flexibility as seen with the emergence of Web services, for example with Microsoft's .NET framework (Schlueter Langdon 2006, 2003b). Specifically, Pääkkönen & Pakkala reveal three major and common steps of data refinement – because of our focus on data refinement, we are explicitly excluding any analysis, analytics and visualization steps: (i) data extraction, loading and pre-processing; (ii) data processing, and (iii) data transformation. This in-depth, case-study based assessment of big data pioneers is corroborated through extensive SLRs: The first study includes 227 articles from peer-reviewed journals extracted from the Scopus database from 1996–2015 (Sivarajah et al. 2017).

It confirms three steps in the data preparation process (again, excluding data analysis, analytics and visualization steps): data intake (acquisition and warehousing), processing (cleansing) and transformation (aggregation and integration; p. 273).

A second, recent study considered 49 articles from three different branches in the literature (Stieglitz et al. 2018): computer science (ACM and IEEE), information systems (AIS), and the social sciences (ScienceDirect). This second SLR yields the addition of data quality as another distinct and common step in the data refinement process (Stieglitz et al. 2018, Fig. 3, p. 165). These four steps as illustrated in Fig. 1 provide the foundation to which we add our observations constructing a real-world data factory. This factory is built by Deutsche Telekom and part of the Telekom Data Intelligence Hub (DIH, Deutsche Telekom 2018). Deutsche Telekom is one of the world's leading integrated telecommunications companies, with some 178 million mobile customers and nearly 50 million fixed-network lines; it operates in more than 50 countries, and generated revenue of 76 billion Euros in the 2018 financial year (Deutsche Telekom 2019). The DIH has been launched as a minimum viable product in late 2018 in Germany at: https://dih.telekom.net/en/. Coming from this practical experience we propose a slightly more granular decomposition of data refinement activities to explicitly recognize issues that have emerged as a critical concern in practice and that require additional data processing steps: data privacy and data sovereignty. Both issues had already surfaced in the SLR by Sivarajah et al. but only as "management challenges" not explicitly as data refinement steps (p. 274). However, since 2018, the General Data Protection Regulation (GDPR) is mandating data privacy protection in the entire European Union, which necessitates additional data refinement steps, such as consent management, anonymization and user data deletion (European Commission 2018). Similarly, the issue of data sovereignty has evolved from a hygiene factor to a key element of a company's business strategy (e.g., Otto 2011) – it even factors into industrial policy of nations: "the question of data sovereignty is key for our competitiveness,' according to Germany's economy minister" (Sorge 2019). And Europe is not alone; in 2018 California became the first U.S. state with a comprehensive consumer privacy law when it enacted the California Consumer Privacy Act of 2018 (CCPA), which becomes effective 2020 (Cal. Civ. Code §§ 1798.100-1798.199). CCPA not only grants residents in California new rights regarding their personal information; more importantly it imposes data protection duties on entities conducting business in California. This matters, because California is a very big market. It is the most populous state in the U.S. and based on its GDP it would rank as the fifth largest economy in the world ahead of Great Britain, France and Italy Fig. 2 (http://www.dof.ca.gov/Forecasting/Economics/Indicators/Gross_State_Product/).

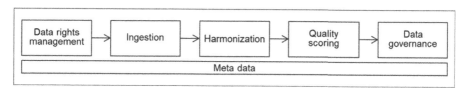

Fig. 2. Extended data factory framework

Legal issues may not be so important from a pure computer science and software engineering perspective. For information systems they certainly matter, because any information system and its architecture would have to correspond with business requirements (Schlueter Langdon 2003a). Therefore, we propose to bookend the data refinement process by data rights management at the beginning to ensure any refinement is compliant with legal requirements in the first place and by data governance at the end to safeguard data sovereignty. Figure 2 illustrates the expanded data factory framework.

In a nutshell raw data rights must be verified before any data can be ingested or harvested (rights, licensing, user consent). Then data ought to be harmonized or properly labeled or tagged for it to be made discoverable through a catalog of categories and search engines (classification). Furthermore, it needs to be scored to provide some indication of quality, because without it any subsequent analytics is pointless – "garbage in, garbage out" (GIGO, quality scoring). Finally, governance mechanisms are required to ensure that data can be exchanged while data sovereignty is maintained for each data provider. For example, in early 2019, Telekom DIH became the first platform to offer data governance controls based on an architecture developed by a consortium of Fraunhofer institutes as illustrated in Fig. 3 (Fraunhofer 2019). Other data factories are emerging. Microsoft is offering "Azure Data Factory" as a feature in its Azure cloud, which is alleviating fears in Europe that hyperscalers are already expanding their dominance beyond data storage (Clemons et al. 2019). In the Azure Data Factory users can "create and schedule data-driven workflows (called pipelines) that can ingest data from disparate … [sources and]… move the data as needed to a centralized location for subsequent processing" (Microsoft 2018). A quick comparison of this description with Fig. 2 reveals that it is so far more narrowly focused on an upstream module, specifically on ingestion. Another "Data Factory," by Datahub, provides open toolkits for data cleaning, modification and validation (Datahub 2019), which - according to Fig. 2 - would be more focused downstream on data classification and quality enhancements.

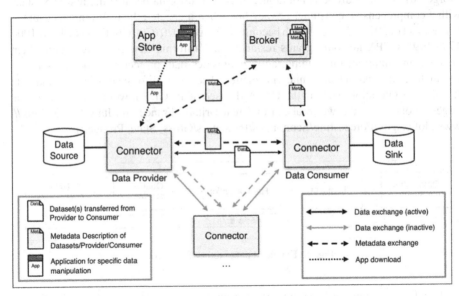

Fig. 3. Data governance architecture (IDSA 2019, p. 59)

2.2 Rights Management

For any product to be marketed and sold ownership rights and licensing rules have to be verified and observed. With data the issue of ownership and rights is complicated. In Germany for example, ownership is typically limited to physical objects ("Sachen, körperliche Gegenstände," § 903 German Civil Code BGB, BMJV 2013). Electronic data is not included. Instead data is protected and rights to it are dealt with in special regulation, such as data protection, copyright and competition laws (for an overview, see Dewenter and Lueth 2018, Chapters 4 and 5). However, since May 2018 complication with personal data rights in the European Union (EU) has been greatly reduced. The EU is now enforcing the General Data Protection Regulation (GDPR, European Commission 2018). GDPR aims to give control to individuals in the EU over their personal data. It has created the legal foundations for a uniform digital single market, which greatly simplifies the regulatory environment for doing business in the EU.

It also has legitimized the business that involves personal data including the user-product interaction data or behavioral data, which has proven to be most valuable for optimization, customization and personalization of customer journeys and the user experience (UX; Crosby and Schlueter Langdon 2014). In a nutshell, in EU countries, any personal data has to be GDPR compliant.

Key requirements include user consent, privacy, fair and transparent processing, limits to storage and use, right to be forgotten, user access and portability and breach reporting duties. GDPR gives individuals control over their personal data and privacy. Unless an individual has provided explicit, informed consent to data processing, personal data may not be processed. Instead of stifling data innovation GDPR has an opposite effect, igniting data industrialization because it legitimizes the use of personal data for business. Being GDPR compliant can avoid embarrassment, erosion of trust and legal trouble. Particularly data leaks involving social media data have accumulated to trigger a public backlash ("surveillance capitalism," Zuboff 2019) to the point that Facebook's CEO Zuckerberg is encouraging introduction of regulation similar to GDPR in the US (Zuckerberg 2019).

For any data factory a first key step will include data rights management to ensure compliance with licensing agreements and GDPR. In case of personal data or data with personally identifiable information an important step is anonymization and pseudonymization. According to Recital 26 of GDPR anonymized data must be stripped of any identifiable information, so that it becomes impossible to generate insights on a discreet individual, even by the party that is responsible for the anonymization. Pseudonymization is less restrictive and requires personal data to be separated into usable data and "additional information" so that "data can no longer be attributed to a specific data subject without use of additional information" (Article 4(5) GDPR). If a data factory is processing particularly personal data at a large scale, a Data Protection Officer (DPO) must be appointed (Article 37 GDPR).

2.3 Ingestion

An early challenge in the data production process is to retrieve all relevant data for an AI application or data product. Often, it involves connecting all required sources and moving the data to a centralized location for subsequent processing (examples of Facebook and Netflix in Pääkkönen and Pakkala 2015). This can become a cumbersome affair as data is typically scattered throughout an enterprise and its supply chain and channel system. In most companies data is stashed away in databases, lots of databases - reflected in the growth of database vendors, like Oracle. Some data has been moved into data warehouses and lately into data lakes (Porter and Heppelmann 2015) - some is held on premise; other data is already stored in public clouds.

In addition to locating and retrieving data from multiple sources it may exist in different formats - and may have to be converted or transformed. For each data type (text, image, audio, video) there are multiple file format options. Examples range from plain text (txt) and csv (comma-separated value) files to video formats (such as MPEG-4), and from open-standard file formats, such as JSON, to proprietary formats (USGS 2019, Oregon State University 2019).

Depending on the application domain and analytics method data may have to be transformed into a form consumable by a particular AI method. This may require additional activities, such as normalizing it and dealing with missing data, corrupted binary data and miss-labeled column descriptions.

2.4 Harmonization

The Cambridge Dictionary defined harmonization as "the act of making different [... elements] suitable for each other, or the result of this (Cambridge Dictionary). For data this includes data classification, which is important for at least two uses cases: (1) discovery and (2) training. In order to reduce the time it takes to find the right information, data has to be labeled or tagged so it can be discovered quickly either through a catalog or search engine. Classification is also required for training data. Raw data has to be labeled and annotated for use in training and validation of machine learning systems. For example, in autonomous driving with footage from onboard cameras, someone must go through each frame and identify people, objects and markings. The deep learning system needs to be told what pedestrians look like and from different angles in different weather conditions in order to generate the computational model of that pattern.

These additional information on raw data is referred to as metadata, which is created along the entire data production pipeline (see Fig. 2). One issue with metadata and labeling in particular is semantic standardization. Different areas and companies introduce their schematic and principles or "vocabulary," which may or may not be compatible with other vendors or domains. These schematics and underlying principles are also referred to as data taxonomies (taxonomy comes from the Greek τάξις, taxis - meaning 'order', 'arrangement'; and νόμος, nomos - 'law' or 'science'; Merriam-Webster).

For example, in areas such as autonomous driving, many specialists have emerged that focus on labeling and annotating raw data for use in training AI. This group includes well established vendors of high definition maps, such as Here and TomTom, as well as startups eager to create their own taxonomy as intellectual property. Recently, two open harmonization efforts have gained traction: OpenDrive to describe entire road networks with respect to all data belonging to the road environment, and OpenScenario to describe the entities acting on or interacting with the road.

On the Web, the The World Wide Web Consortium (W3C), the international community that develops open Web standards, has published a framework and recommendations on web annotation complete with a model (describes underlying abstract data structure), vocabulary (which underpins the model) and protocol (HTTP API for publishing, syndicating, and distributing Web Annotations) (https://www.w3.org/annotation/).

2.5 Quality Scoring

Quality scoring is a well-established business - but not with data. As consumers, most of us are probably familiar with Consumer Reports in the US ("Stiftung Warentest" in

Germany), which is testing and rating consumer products; car buyers are likely checking J.D. Power's quality scores from the vendor's Initial Quality Study (IQS, problems after 3 months) and Vehicle Dependability Study (VDS, problems after 3 years); home buyers are worried about credit scores (FICO in the US, SCHUFA in Germany) and familiar with credit scoring agencies, such as Equifax; and finally, bond investors watch ratings of creditworthiness of corporate bonds ratings from specialists, like Moody's (Aa1) and Standard & Poor's (AA+).

As of 2019, for data, similar quality scoring solutions are missing. On one hand, this is a surprise because the importance of data quality seems to be unequivocally acknowledged, and therefore, scoring is recognized as a core data factory module (as illustrated in Fig. 2). On the other hand, data may be too complex with varying flavors across domains and industries to warrant an easy solution, which in turn presents a business opportunity.

The data science community confirms the old adage of "garbage in, garbage out" (GIGO): "Dirty data" is seen as the most common problem for workers in data science according to a survey with 16,000 responses on Kaggle (Kaggle 2017). With data analytics all insights are extracted from inside the data. Therefore, it is imperative to ensure that any raw data used has the information required for insights in it. One analogy is iron ore: For iron one would need rocks so rich with iron oxides that metallic iron can be extracted. Without iron oxide in it a rock would simply be a rock not iron ore.

Yet, despite the importance of data quality much work remains custom, hand-made and qualitative. The literature is using concepts, such as the "3 Vs" of volume, velocity and variety (McAfee and Brynjolfsson 2012); and more Vs are being added, like variability and value (e.g., Yin and Kaynak 2015). However, from an operational perspective, from an analytics application point of view, the Vs have remained conceptual and qualitative. The Vs may be useful for a first assessment, maybe for a pre-test, a first triage type data selection. However, in order to gauge outcomes in terms of performance, to estimate the likelihood of effects (x improves y), the size of effects (x improves y by a lot) and significance (improvements are real not random), the Vs have too little information in them. For example, consider traffic data for a routing app or parking data for a parking app: How fresh is the data? How frequently updated? Or consider time series data: How long and granular is it? Length of the overall observation period (10 years as opposed to 1 year), and for any period how dense is the data (one year's worth of data in monthly, daily, hourly intervals?). Our survey confirms the quality rating opportunity: While quality is preferred over quantity (How to spend the next US$1? Quality: 82%, quantity: 18%; n = 65), no obvious quality indicator is emerging (volume: 3%, freshness: 30%, format: 34%, source: 31%, license type: 2%, n = 64).

2.6 Governance

Many AI applications, such as predictive maintenance or autonomous driving, can require more data than what is available within a single department and company. Creating data pools across companies would be an advantage (IDSA 2019, Fig. 2.3, p. 15). For example, pooling all data of a particular machine type across all installations (horizontal pooling) would create a rich dataset for anomaly detection and its root-cause analysis.

Another use case is pooling data vertically, across the participants along an entire supply chain or channel system in order to better estimate arrival times or ensure proper end-to-end temperature treatment of shipments, for example. In both situations, horizontal and vertical pools, outcomes would be best if most participants were to contribute. However, so far, few companies have been willing to engage in this type of data sharing. On one hand, data is increasingly seen as a strategic advantage (the value aspect of "data is the new oil"), and therefore, held closely and protected. On the other hand, more sensor data will only increase data pooling benefits. What has been missing are exchange options with data governance mechanisms that strike a balance between the need to protect one's data and share it with others (Otto et al. 2016; IDSA 2018a, 2018b).

Such data governance solutions are emerging. An important example is the reference architecture model (RAM) of the International Data Spaces Association (IDSA 2019). IDSA is an association of industry participants, created to promote data governance architecture solutions based on research conducted by German Fraunhofer Institute with funding from the German government (Fraunhofer 2015).

Members include automakers like Volkswagen, suppliers like Bosch, and traditional information technology specialists like IBM.

The core element of IDSA RAM is a "connector." It ensures that data rights can be governed. Figure 3 illustrates the role of this element in the data flow between source (data provider) and sink (data consumer). With a connector any data package or product can be "wrapped up" in instructions and rules for use. Technically, it is a dedicated software component allowing participants to exchange, share and process data such that the data sovereignty of the data owner can be guaranteed.

Depending on the type of configuration, the connector's tamper-proof runtime can host a variety of system services including secure bidirectional communication, enforcement of content usage policies (e.g., expiration times and mandatory deletion of data), system monitoring, and logging of content transactions for clearing purposes.

As illustrated in Fig. 3, the functional range of a connector may be extended by (a) custom data apps, such as data visualization, provided in an app store and (b) a broker function to allow for product listings, such as a marketplace menu, and clearing services. A first connector implementation has been certified by IDSA for Deutsche Telekom's Data Intelligence Hub (Fraunhofer 2019).

3 Concluding Comments

Data and data analytics are seen as the next big business opportunity. Yet, today, data analytics is handmade and the overwhelming share of the time budget of a data analytics project is spent on refining data. In order to boost productivity and to prepare for even more IoT data while minimizing the risk of non-compliance with regulation, a "data factory" is required for data productization in an automated manner. Building on a foundation in the literature we have added bookends on data rights management and data governance in response to emerging data regulation. A data factory can be internal or external: It can be operated internally within the IT function (e.g., under a Chief Information Officer, CIO) or outside of it (e.g., under a Chief Marketing Officer, CMO), or it can be a separate, standalone business entirely. First standalone data factory service offerings have already arrived with large enterprises, for example in Microsoft's Azure cloud and in the Telekom Data Intelligence Hub. Internal data factories may provide a way forward to extract value from data lakes and convert cost into business advantage by creating data products for internal operations and applications, such as anomaly detection, or into top-line growth with the sale of data products to third parties. Finally, combining a data factory with a data exchange may be an elegant way for large multi-divisional companies to quickly enable and promote a data-centric organization across functional or departmental silos. It's a classic: Top down push versus bottom up pull. Instead of having to define details upfront and top down, such as data product types and quality standards, and to enforce cooperation across silos, the market forces of an exchange would weed out lemon products and make data attributes and quality transparent.

References

Akerlof, G.A.: The market for 'lemons': quality uncertainty and the market mechanism. Q. J. Econo. **84**(3), 488–500 (1970)

American Marketing Association The 2018 AMA gold top 50 report (2018). https://www.ama.org/marketing-news/the-2018-ama-gold-top-50-report/

BMJV, Federal Ministry of Justice and Consumer Protection German Civil Code BGB (2013). http://www.gesetze-im-internet.de/englisch_bgb/index.html

Clark, K.B., Fujimoto, T.: Product Development Performance: Strategy, Organization, and Management in the World Auto Industry. In: Harvard Business School Press: Boston, MA (1991)

Clemons, E.K., Krcmar, H., Hermes, S., Choi, J.: American domination of the net: a preliminary ethnographic exploration of causes, economic implications for Europe, and future prospects. In: 52nd Hawaii International Conference on System Sciences (HICSS) (2019). https://doi.org/10.24251/hicss.2019.737

Crosby, L., Langdon, C.S.: Data as a product to be managed. Mark. News Am. Mark. Assoc. (2019). https://www.ama.org/marketing-news/data-is-a-product

Crosby, L, Langdon, C.S.: Technology personified. Mark. News Am. Mark. Assoc. 18–19 (2014)

Deutsche Telekom At a glance (2019). https://www.telekom.com/en/company/at-a-glance

Deutsche Telekom Creating value: Deutsche Telekom makes data available as a raw material. Press Release (2018). https://www.telekom.com/en/media/media-information/archive/deutsche-telekom-makes-data-available-as-a-raw-material-542866

Dewenter, R., Lueth, H.: Datenhandel und Plattformen. Gutachten. In: ABIDA – Assessing Big Data, German Federal Ministry of Education and Research (2018)

European Commission General Data Protection Regulation (2018). https://ec.europa.eu/commission/priorities/justice-and-fundamental-rights/data-protection/2018-reform-eu-data-protection-rules_en

Fraunhofer: Hannover Tradefair 2019, 'International Data Space'-Architecture implemented in first digital ecosystems. Fraunhofer Institute for Software and Systems Engineering, Press Release (2019). https://www.isst.fraunhofer.de/en/events/InternatData_Space-Architecture_implemented_in_first_digital_ecosystems.html

Fraunhofer: Fraunhofer initiative for secure data space launched. Fraunhofer Society for the Advancement of Applied Research, Press Release (2015). https://www.fraunhofer.de/en/press/research-news/2015/september/Fraunhofer-initiative-for-secure-data-space-launched.html

Gartner: Gartner top 10 strategic technology trends for 2019 (2018a). https://www.gartner.com/smarterwithgartner/gartner-top-10-strategic-technology-trends-for-2019/

Gartner: Gartner forecasts worldwide public cloud revenue to grow 17.3 percent in 2019 (2018b). https://www.gartner.com/en/newsroom/press-releases/2018-09-12-gartner-forecasts-worldwide-public-cloud-revenue-to-grow-17-percent-in-2019

Glassberg Sands, E.: How to build great data products. In: Harvard Business Review (2018). https://hbr.org/2018/10/how-to-build-great-data-products

IDSA Reference architecture model. International Data Spaces Association, Version 3.0 (2019). https://www.internationaldataspaces.org/info-package/

IDSA Sharing data while keeping data ownership, the potential of IDS for the data economy. International Data Spaces Association, White Paper (October) (2018a). https://www.internationaldataspaces.org/publications/sharing-data-while-keeping-data-ownership-the-potential-of-ids-for-the-data-economy/

IDSA Jointly paving the way for a data driven digitisation of European industry. International Data Spaces Association, White Paper Version 1.0 (October) (2018b). https://www.internationaldataspaces.org/publications/strategic-paper-for-europe-idsa/

Kaggle The state of data science & machine learning (2017). https://www.kaggle.com/surveys/2017

Krzanich, B.: Data is the new oil in the future of automated driving. Intel Newsroom (2016). https://newsroom.intel.com/editorials/krzanich-the-future-of-automated-driving/#gs.6pqyxh

McAffee, A., Brynjolfsson, E.: Big data the management revolution. Harvard Bus. Rev. **90**(10), 60–68 (2012)

McKinsey Global Institute Notes from the AI frontier - insights from hundreds of use cases. McKinsey & Company (2018). https://www.mckinsey.com/featured-insights/artificial-intelligence/notes-from-the-ai-frontier-applications-and-value-of-deep-learning

McKinsey & Company Monetizing car data. Advanced Industries Report (2016). https://www.mckinsey.com/industries/automotive-and-assembly/our-insights/monetizing-car-data

Melendez, S.: A landmark Vermont law nudges over 120 data brokers out of the shadows. Fast Company (2019). https://www.fastcompany.com/90302036/over-120-data-brokers-inch-out-of-the-shadows-under-landmark-vermont-law

Microsoft Introduction to azure data factory (2018). https://docs.microsoft.com/en-us/azure/data-factory/introduction

Miller, R.: AWS and Microsoft reap most of the benefits of expanding cloud market. In: Techcrunch (2019). https://techcrunch.com/2019/02/01/aws-and-microsoft-reap-most-of-the-benefits-of-expanding-cloud-market/

Oregon State University Research data services: Data types & file formats (2019). https://guides.library.oregonstate.edu/research-data-services/data-management-types-formats

Otto, B., Juerjens, J., Schon, J., Auer, S., Menz, N., Wenzel, S., Cirullies, J.: Industrial data space - digital sovereignty over data. Fraunhofer Society for the Advancement of Applied Research (Working Paper) (2016). https://www.fraunhofer.de/content/dam/zv/en/fields-of-research/industrial-data-space/whitepaper-industrial-data-space-eng.pdf

Otto, B.: Organizing data governance: findings from the telecommunications industry and consequences for large service providers. Commun. AIS **29**(1), 45–66 (2011)

Palmer, M.: Data is the new oil. CMO News, ANA - Association of National Advertisers (2006). https://ana.blogs.com/maestros/2006/11/data_is_the_new.html

Pekka Pääkkönen, P., Pakkala, D.: Reference architecture and classification of technologies, products and services for big data systems. Big Data Res. **2**, 166–186 (2015)

Porter, M.E., Heppelmann, J.E.: How smart, connected products are transforming companies. Harvard Bus. Rev. (2015). https://hbr.org/2015/10/how-smart-connected-products-are-transforming-companies

Press, G.: Cleaning big data: Most time-consuming, least enjoyable data science task, Survey Says. Forbes (2016)

Schlueter Langdon, C.: Designing information systems capabilities to create business value: a theoretical conceptualization of the role of flexibility and integration. J. Data. Manage. **17**(3), 1–18 (2006)

Schlueter Langdon, C.: Information systems architecture styles and business interaction patterns: toward theoretic correspondence. J. Inf. Syst. E-Bus. **1**(3), 283–304 (2003a)

Schlueter Langdon, C.: The state of web services. IEEE Comput. **36**(7), 93–95 (2003b)

Sivarajah, U., Kamal, M.M., Irani, Z., Weerakkody, V.: Critical analysis of big data challenges and analytical methods. J. Bus. Res. **70**, 263–286 (2017)

Sorge, P.: Germany backs european cloud project to avoid dependence on US technology. Wall Street J. (2019). https://www.wsj.com/articles/BT-CO-20190924-705704

Stieglitz, S., Mirbabayea, M., Rossa, B., Neubergerb, C.: Social media analytics – challenges in topic discovery, data collection, and data preparation. Int. J. Inf. Manage. **39**, 156–168 (2018)

Stern, D., Herbrich, R., Graepel, T.: Matchbox: large scale bayesian recommendations. In: Proceedings of the 18th International World Wide Web Conference (2009). https://www.microsoft.com/en-us/research/publication/matchbox-large-scale-bayesian-recommendations/?from=http%3A%2F%2Fresearch.microsoft.com%2Fpubs%2F79460%2Fwww09.pdf

USGS Data & file formats. United States Geological Survey (2019). https://www.usgs.gov/products/data-and-tools/data-management/data-file-formats

Vermont Office of the Attorney General Guidance on Vermont's Act 171 of 2018 Data Broker Regulation (2018). https://ago.vermont.gov/wp-content/uploads/2018/12/2018-12-11-VT-Data-Broker-Regulation-Guidance.pdf

Vollenweider, M.: Mind + Machine: A decision model for optimization and implementing analytics. John Wiley & Sons: Hoboken, NJ (2016)

Wall, M.: Tech trends 2019: 'The end of truth as we know it?' In: BBC (2019). https://www.bbc.com/news/business-46745742

Walter, S.M., Böhmann, T., Krcmar, H.: Industrialisierung der IT — Grundlagen, Merkmale und Ausprägungen eines Trends. HMD Praxis der Wirtschaftsinformatik **44**(4), 6–16 (2007). https://doi.org/10.1007/BF03340302

Womack, J., Jones, D., Roos, D.: The Machine that Changed the World: The Story of Lean Production. Free Press, Simon & Schuster, New York, NY (1990)

Yin, S., Kaynak, O.: Big data for modern industry: challenges and trends. Proc. IEEE **103**(2), 143–146 (2015)

Zuboff, S.: The Age of surveillance capitalism: the fight for a human future at the new frontier of power. New York, NY, Hachett Book Group (2019)

Zuckerberg, M.: The facts about facebook - We need your information for operation and security, but you control whether we use it for advertising. Wall Street J. (2019). https://www.wsj.com/articles/the-facts-about-facebook-11548374613

An Empirical Investigation of Analytics Capabilities in the Supply Chain

Thiagarajan Ramakrishnan[1], Abhishek Kathuria[2(✉)] [iD], and Jiban Khuntia[3]

[1] Prairie View A&M University, 700 University Drive, Prairie View, TX 77446, USA
ram@pvamu.edu
[2] Indian School of Business, Hyderabad, Telangana, India
[3] University of Colorado Denver, 1475 Lawrence Street, Denver, CO, USA

Abstract. This study conceptually develops the construct of Supply Chain Analytics Capability. Preliminary analysis of survey data collected from more than 100 firms in India supports several hypotheses relating supply chain analytics architecture modularity and decentralized governance, in a moderating manner, with Supply Chain Analytics Capability. The capability creation path model suggested in this study establishes the antecedents of Supply Chain Analytics Capability and is helpful to firms seeking to develop such a capability.

Keywords: Supply chain · Analytics · Capabilities · Partial least squares · India

1 Introduction

Despite making increasing investments in Business Intelligence and Analytics (BI&A), many firms struggle to strategically leverage the data handled by these systems. Their challenge has been predominantly to use analytics in a manner that is aligned to business strategies and value-chain activities. While intelligence and analytics have been used to scan and profile customers, their use for supply chain relevant value chain activities is sparsely quoted.

Implementation of analytics capabilities for supply chain management (SCM) is important to enhance a firm's competitive advantage by improving supplier or customer relations, attaining operational flexibility, and/or by lowering production costs (Sahay and Ranjan 2008). Innovations in SCM, such as location-based inventory system and use of tags to track moving items in the supply chain, paired with BI&A have helped organizations to achieve better information integration and coordination (Rai et al. 2006). These have the potential to provide better vendor and inventory management, logistical efficiencies, higher supply chain profit and improved product innovation.

Using BI&A for supply chain operations is not enough when it comes to applications of analytics in supply chain area (Ramakrishnan et al. 2020). While basic operations oriented analytic capabilities form the comprehensive analytic capabilities required for improving supply chain activities within an organization, in general, additional or complex analytical capabilities along the supply chain provide the orientation to predict and prepare for future uncertainties. Critically, what and where analytics applications should

© Springer Nature Switzerland AG 2020
K. R. Lang et al. (Eds.): WeB 2019, LNBIP 403, pp. 56–63, 2020.
https://doi.org/10.1007/978-3-030-67781-7_6

be implemented along the supply chain are decisions that need to be taken by the focal firm not independently, but after engaging with supply chain partners.

In this study, we approach the challenge of leveraging analytics capabilities along the supply chain. We do that in three ways: (1) using a construct to capture *Supply Chain Analytics Capability*, (2) using an additional construct to frame and capture standardized and loosely coupled modular analytics architecture for supply chain, and (3) exploring what governance conditions for analytics capabilities among supply chain partners have the greatest influence on *Supply Chain Analytics Capability*. We propose a conceptual model linking architecture modularity and governance along the supply chain to analytical capabilities. Partial least square analysis of primary survey data collected from more than 100 firms in India supports the hypotheses.

2   Literature Review

2.1   Supply Chain Analytics Capability

The main objective of BI&A is to integrate and analyze data from disparate sources in a timely manner (Ramakrishnan et al. 2012; Ramakrishnan et al. 2016; Ramakrishnan et al. 2018; Ramakrishnan et al. 2020). We draw on the SCOR (supply chain operations reference) model developed by the *Supply Chain Council* (and now run by APICS) to examine the different areas of supply chain where analytics can be used. The SCOR model provides four major activities of plan, make, source, and deliver included in supply chain management. Accordingly, we define *Supply Chain Analytics Capability* as the degree to which various techniques are used to analyze data for improving supply chain activities and relationships. It consists of *Planning Analytics Capability*, *Making Analytics Capability*, *Sourcing Analytics Capability*, and *Delivery Analytics Capability* based on the four activities described for supply chain management as prescribed by the *Supply Chain Council*.

Planning Analytics Capability refers to the use of analytics during the planning process of supply chain management. This capability helps firms predict market trends regarding their products and services. *Sourcing Analytics Capability* focuses on procurement of raw materials and in evaluating suppliers and thereby improving supplier selection. *Making Analytics Capability* is used in different areas such as identifying irregularities in production process, predicting machinery failures, scheduling the production inventory items with regards to time, belt, and batch. Thus, *Making Analytics Capability* improves the operation process of the value chain. *Delivery Analytics Capability* improves the out-bound logistics of the value chain. Thus, this capability improves the efficiency of bringing the product to the market.

In summary, analytics capabilities of planning, sourcing, making, and delivery along the supply chain form the four dimensions of *Supply Chain Analytics Capability*. An increase in either of these dimensions may increase the overall *Supply Chain Analytics Capability* without affecting the other dimensions. Therefore, *Supply Chain Analytics Capability* is operationalized as a second order formative construct (Fig. 1).

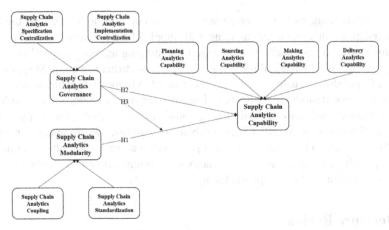

Fig. 1. Conceptual model

2.2 Supply Chain Analytics Modularity

Supply Chain Analytics includes a collection of tools for analyzing, manipulating, and mining the data in the data warehouse (Ramakrishnan et al. 2020). *Supply Chain Analytics Modularity* is defined as the arrangement through which different process and data from different systems for supply chain can be integrated within an organization (Tiwana and Konsynski 2010). *Supply Chain Analytics Modularity* can be achieved through *Supply Chain Analytics Coupling* and *Supply Chain Analytics Standardization*. *Supply Chain Analytics Coupling* refers to the degree to which supply chain wide standards and policies pre-specify how analytics application in supply chain connect and interoperate. *Supply Chain Analytics Standardization* refers to the design of supply chain analytics applications in such a way that changes made to one application by a supply chain partner does not affect the behaviors the applications of others. *Supply Chain Analytics Modularity* is conceptualized as a second order formative construct.

2.3 Supply Chain Analytics Governance

We examined the literature on IT Governance to extrapolate *Supply Chain Analytics Governance*. Governance is about providing authority to granting decision rights, defining expectations, and ensuring performance (Tiwana and Konsynski 2010). We define *Supply Chain Analytics Governance* as the sharing of analytics related decision-making authority between the focal firm and supply chain partners. We use the two categories of specification and implementation to conceptualize *Supply Chain Analytics Governance*. *Supply Chain Analytics Specification Centralization* and *Supply Chain Analytics Implementation Centralization* form two dimensions of *Supply Chain Analytics Governance*, which is conceptualized as a second order formative construct.

3 Conceptual Model and Hypotheses Development

3.1 Hypotheses Development

Supply Chain Analytics Capability is formed of the different analytical capabilities that are important to leverage the different activities within the value chain. These may require the integration of different analytical applications. The ability to add new or modified analytics components to the existing stack of analytics applications without impacting them is critical. Thus, a highly modular architecture will enable *Supply Chain Analytics Capability.*

Hypothesis (H1): There is a positive relationship between Supply Chain Analytics Modularity and Supply Chain Analytics Capability.

Supply Chain Analytics Governance deals with the relationship between the focal firm and its supply chain partners in deciding and implementing analytical applications for supply chain management. Like line functions within organizations, supply chain partners are more familiar to their own operational needs (Sambamurthy and Zmud 2000). A higher supply chain analytics governance implies a higher degree of shared responsibilities between the focal firm and its partners with regards to supply chain management and thus more effective and efficient use of supply chain analytics. Thus, better *Supply Chain Analytics Governance* will improve the *Supply Chain Analytics Capability* of the firm.

Hypothesis (H2): There is a positive relationship between Supply Chain Analytics Governance and Supply Chain Analytics Capability.

Even though modularity of supply chain analytics architecture improves *Supply Chain Analytics Capability,* this relationship is stronger when complemented with *Supply Chain Analytics Governance.* Further, in order to take advantage of the four first-order *Supply Chain Analytics Capabilities* and implement effective supply chain analytics it is important to incorporate high level of independence among these applications while developing the modular architecture. Thus, increasing coupling when designing the supply chain modular architecture will help in enhancing *Supply Chain Analytics Capabilities.* At the same time, it is also important to have highly standard interfaces that allows each of these applications to communicate with each other and with other systems that collect data regarding the supply chain activities. Thus, an increasing *Supply Chain Analytics Modularity* in conjunction with *Supply Chain Analytics Governance* will result in an improved *Supply Chain Analytics Capability.*

Hypothesis (H3): Supply Chain Analytics Modularity in conjunction with Supply Chain Analytics Governance is positively associated with Supply Chain Analytics Capability.

4   Research Methodology

We conduct a cross-sectional matched-pair field survey of manufacturing organizations in India to test our research model. India is the world's third largest and fastest growing major economy, with manufacturing contributing 17% of its gross domestic product (Celly et al. 2016). Thus, it is an increasing context for emergent research (e.g., Kathuria et al. 2020; Kathuria et al. 2018a; Kathuria et al. 2018b; Kathuria et al. 2016; Khuntia et al. 2019; Ramakrishnan et al. 2016), especially given that India is an exemplar of a G.R.E.A.T. domain (growing, rural, eastern, aspirational, and transitional), which together account for a significant proportion of world population and economic output, yet suffer from lack of research attention (Karhade and Kathuria 2020). To minimize confounding factors, we developed a sample frame of 1000+ firms that were in western India were used, following industry directories (Kathuria et al. 2018a provide detailed protocols for primary data collection in India; Kathuria et al. 2018b; Kathuria et al. 2010; Khuntia et al. 2014; Khuntia et al. 2019). We utilized existing scales where available or else developed new multi-item scales for measuring each of our first or second order constructs. The questions for the survey were localized by employing the *back-translation method* (Kathuria et al. 2018a; Khuntia et al. 2019). Common strategies for survey administration was followed, along with a *dual online-offline mode*, which is crucial to primary data collection efforts in India (see (Kathuria et al. 2018a; Khuntia et al. 2019) for details of the method). A final sample of 127 firms was used for analysis.

5   Data Analysis and Results

5.1   Assessment of Measurement Model

We evaluated internal consistency and reliability by assessing composite reliability scores and Cronbach's α. Both variables exhibited sufficiently high reliability, with satisfactory composite reliability at 0.92 and 0.89 and Cronbach's α at 0.87 and 0.84 respectively. Convergent validity was assessed by evaluating *average variances extracted* and outer loadings. Discriminant validity was evaluated through cross-loading analysis and the *heterotrait-monotrait ratio*. Third, we assessed the first- and second-order formative constructs. We assessed convergent and discriminant validity by evaluating the weight, sign, and magnitude of items (Henseler et al. 2015; Petter et al. 2007). Other standard measurement model relevant evaluations were deployed.

5.2   Assessment of Structural Model

We assessed the structural model by using SmartPLS 3.0 (Hair Jr et al. 2016). We initially executed the analysis without the moderating paths and added the moderations in subsequent analysis (Henseler and Fassott 2010). Our results indicate that *Supply Chain Analytics Modularity* had a positive relationship with *Supply Chain Analytics Capability* ($\beta = 0.48$, $t = 7.39$, $p < 0.01$), supporting Hypothesis 1. *Supply Chain Analytics Governance* also had a positive relationship with *Supply Chain Analytics Capability* ($\beta = 0.50$, $t = 7.73$, $p < 0.01$), supporting Hypothesis 2. Further, *Supply*

Chain Analytics Governance in conjunction with Supply Chain Analytics Modularity had positive relationships with *Supply Chain Analytics Capability* ($\beta = 0.06$, $t = 1.49$, $p < 0.10$), supporting Hypothesis 3.

6   Discussion

6.1   Key Findings

To summarize our results, we find that to build a robust *Supply Chain Analytics Capability*, modularity and governance are two important precursors. In addition, governance and modularity attributes must align with each other to create higher level of analytics capabilities.

6.2   Implications for Theory

The theoretical implications of this study are significant. This study takes a 'capability creation' path model (Kathuria et al. 2018) for *Supply Chain Analytics* and establishes how an integrated analytics capability can be created along the supply chain for effective performance (Wright et al. 2005). This is a major contribution of this study.

Second, unlike earlier studies that focus on supply chain information integration, we demonstrate that the antecedents of supply chain analytics are visible at finer levels of detail by making conceptual and empirical distinctions through the direct and moderating roles of analytics architecture and governance.

Third, study deepens scholarly understanding of the SCM and IS issues that lead to competitive advantage in emerging markets, addressing the call to study the SCM issues in non-US in particular, and emerging contexts (Fawcett and Waller 2015) and calls to examine IS issues in G.R.E.A.T. contexts in general (Karhade and Kathuria 2020). The experiences of market leaders trying to establish businesses in India point to the fact that there are idiosyncrasies in the emerging context that need deeper investigation (Kathuria et al. 2020; Kathuria and Konsynski 2012). This is consistent with earlier research that suggests that only specific types of capabilities are transferrable from developed to emerging markets and G.R.E.A.T. domains (Karhade and Kathuria 2020; Kathuria and Karhade 2019; Kathuria et al. 2018a; Kathuria et al. 2018b; Khuntia et al. 2019).

Finally, while integration in supply chains has generated substantial attention in both theoretical and empirical studies, the literature provides limited evidence about the role of analytics in this integration process. This issue is highly relevant in emerging markets, as reflected in the context of this study, as compared with developed markets (Khuntia et al. 2019). However, developed markets are not without supply chain integration challenges, as many firms are suffering from chronic supply chain management disorders, even after been highly successful companies for more than several decades. In this context, this study fulfills a gap by examining the antecedents of a robust supply chain analytics capability.

7 Conclusion

To summarize, this study proposes and tests a model for exploring the antecedents of analytics capabilities in supply chains, which result in an integrated *Supply Chain Analytics Capability.*

References

Celly, N., Kathuria, A., and Subramanian, V.: Overview of Indian multinationals. Emerg. Indian Multi. Strateg. Players Multipolar World. Oxford University Press, London (2016)

Fawcett, S.E., Waller, M.A.: Designing the supply chain for success at the bottom of the pyramid. J. Bus. Logistics **36**(3), 233–239 (2015)

Hair Jr, J.F., Hult, G.T.M., Ringle, C., Sarstedt, M.: A Primer on Partial Least Squares Structural Equation Modeling (PLS-SEM). Sage Publications, Thousand Oaks (2016)

Henseler, J., Fassott, G.: Testing moderating effects in PLS path models: an illustration of available procedures. Handbook of Partial Least Squares, pp. 713–735. Springer, Berlin (2010). https://doi.org/10.1007/978-3-540-32827-8_31

Henseler, J., Ringle, C.M., Sarstedt, M.: A new criterion for assessing discriminant validity in variance-based structural equation modeling. J. Acad. Mark. Sci. **43**(1), 115–135 (2014). https://doi.org/10.1007/s11747-014-0403-8

Karhade, P.P., Kathuria, A.: Missing impact of ratings on platform participation in India: a call for research in G. R. E. A. T domains. Commun. Assoc. Inf. Syst. **47**(1), 19 (2020)

Kathuria, A., Karhade, P.: You are not you when you are hungry: machine learning investigation of impact of ratings on ratee decision making. In: Xu, J.J., et al. (eds.) WEB 2018. LNBIP, vol. 357, pp. 151–161. Springer, Cham (2019). https://doi.org/10.1007/978-3-030-22784-5_15

Kathuria, A., Karhade, P.P., Konsynski, B.R.: In the realm of hungry ghosts: multi-level theory for supplier participation on digital platforms. J. Manage. Inform. Syst. **37**(2), 396–430 (2020)

Kathuria, A., Konsynski, B.R.: Juggling paradoxical strategies: the emergent role of IT capabilities. In: Proceedings of the International Conference on Information Systems, Association of Information Systems, Orlando (2012)

Kathuria, A., Mann, A., Khuntia, J., Saldanha, T.J.V., Kauffman, R.J.: A strategic value appropriation path for cloud computing. J. Manage. Inform. Syst. **35**(3), 740–775 (2018a)

Kathuria, A., Saldanha, T.J.V., Khuntia, J., Andrade Rojas, M.G.: How information management capability affects innovation capability and firm performance under turbulence: evidence from India. In: Proceedings of the International Conference on Information Systems, Association of Information Systems, Dublin (2016)

Kathuria, R., Kathuria, N.N., Kathuria, A.: Mutually supportive or trade-offs: an analysis of competitive priorities in the emerging economy of India. J. High Technol. Manage. Res. **29**(2), 227–236 (2018b)

Kathuria, R., Porth, S.J., Kathuria, N.N., Kohli, T.K.: Competitive priorities and strategic consensus in emerging economies: evidence from India. Int. J. Oper. Product. Manage. **30**(8), 879–896 (2010)

Khuntia, J., Saldanha, T.J.V., Kathuria, A., Konsynski, B.R.: Benefits of IT-enabled flexibilities for foreign versus local firms in emerging economies. J. Manage. Inform. Syst. **36**(3), 855–892 (2019)

Petter, S., Straub, D., Rai, A.: Specifying formative constructs in information systems research. MIS Q. **31**(4), 623–656 (2007)

Rai, A., Patnayakuni, R., Seth, N.: Firm performance impacts of digitally enabled supply chain integration capabilities. MIS Q. **30**(2), 225–246 (2006)

Ramakrishnan, T., Jones, M.C., Sidorova, A.: Factors influencing business intelligence (BI) data collection strategies: an empirical investigation. Decis. Support Syst. **52**(2), 486–496 (2012)

Ramakrishnan, T., Kathuria, A., Saldanha, T.J.V.: Business intelligence and analytics (BI&A) capabilities in healthcare. In: Theory and Practice of Business Intelligence in Healthcare, pp. 1–17. IGI Global, Pennsylvania (2020)

Ramakrishnan, T., Khuntia, J., Kathuria, A., Saldanha, T.J.V.: Business intelligence capabilities and effectiveness: an integrative model. In: Proceedings of the 49th Hawaii International Conference on System Sciences, Koloa, Hawaii. IEEE (2016)

Ramakrishnan, T., Khuntia, J., Kathuria, A., Saldanha, T.J.V.: Business intelligence capabilities. In: Analytics and Data Science: Advances in Research and Pedagogy, Springer International Publishing, Cham, 15–27 (2018)

Ramakrishnan, T., Khuntia, J., Kathuria, A., Saldanha, T.J.V.: An integrated model of business intelligence & analytics capabilities and organizational performance. Commun. Assoc. Inf. Syst. **46**(1), 31 (2020)

Sahay, B., Ranjan, J.: Real time business intelligence in supply chain analytics. Inf. Manage. Comput. Secur. **16**(1), 28–48 (2008)

Sambamurthy, V., Zmud, R.W.: Research commentary: the organizing logic for an enterprise's IT activities in the digital era—a prognosis of practice and a call for research. Inf. Syst. Res. **11**(2), 105–114 (2000)

Tiwana, A., Konsynski, B.R.: Complementarities between organizational IT architecture and governance structure. Inf. Syst. Res. **21**(2), 288–304 (2010)

Wright, M., Filatotchev, I., Hoskisson, R.E., Peng, M.W.: Strategy research in emerging economies: challenging the conventional wisdom. J. Manage. Stud. **42**(1), 1–33 (2005)

Finding Real-Life Doppelgangers on Campus with MTCNN and CNN-Based Face Recognition

Jingjing Ye[1] and Yilu Zhou[2(✉)]

[1] ShanghaiTech University, Shanghai 201210, China
yejj@shanghaitech.edu.cn
[2] Fordham University, New York, NY 10023, USA
yzhou62@fordham.edu

Abstract. Face recognition has been widely used in areas such as security informatics, forensic investigation, customer tracking and mobile payment. This project is inspired by a photography artwork by Francois Brunelle, where he spent 12 years tracking people who are completely strangers but lookalikes, or doppelgangers. We aim to use face recognition techniques to mine doppelgangers on a school campus. We developed a face processing system which includes four steps, face detection, image processing (alignment and cropping), feature extraction and classification. We trained Multi-task Cascaded Convolutional Networks (MTCNN) and traditional CNNs with joined softmax loss and Center Loss on the Caffe framework. Finally, cosine similarity is used to detect similar faces. By exhibiting the results, we demonstrate the potential to adopt CV technology in art-related domains, in this case mimicking a photographer's human eyes. This project provides an example for cross-disciplinary study between art and technology and will inspire researchers from both domains to establish further collaboration channels.

Keywords: Doppelganger · Face recognition · Face detection · MTCNN · CNN · Softmax loss · Center loss · Cosine similarity

1 Introduction

Face recognition technology has shown its wide applications in security informatics, forensic investigation, customer tracking and mobile payment. The general field of computer vision has shown much application in the art domain. For example, in 2016 a team of technologists produced a 3D-printed painting in the style of Dutch master Rembrandt (https://www.nextrembrandt.com/) with AI algorithm and a dataset of 15,000 portraits from the 14th and 20th centuries. It was sold for $432,500 (Time 2018). Such activities have inspired further interest among computer scientists to look for new applications of computer algorithms in the art world. It also encourages artists to seek new technology as a new way to present art. In the specific area of face recognition, we are inspired by photographer François Brunelle's work on "Me, Myself and I". He spent 12 years running around in many countries and cities to find couples who are complete strangers but look like each other, or doppelgangers (Chase Jarvis Photography 2018).

© Springer Nature Switzerland AG 2020
K. R. Lang et al. (Eds.): WeB 2019, LNBIP 403, pp. 64–76, 2020.
https://doi.org/10.1007/978-3-030-67781-7_7

His art exhibition attracted many people. Many people wonder, is there another me in the world? Figure 1 shows some doppelgangers he photographed for the exhibition. Following the work of François Brunelle, this study aims to mine doppelgangers on a university campus using face detection and face recognition algorithms and hope to inspire more cross-disciplinary studies with creative art domain.

The contributions of this study are follows. Firstly, we demonstrate the potential of using technology in creative art. We show that technology can be more accurate in capturing similar faces and eliminate human visual bias affected by skin colors and gender. Secondly, the result of this study will be displayed as an exhibition on campus mimicking Brunel's exhibition. Further interactive activities will allow visitors to find their similar face on campus and test their face similarity with friends. With an introduction of recent development in Deep Learning and Image Processing technology in the exhibition, it will inspire students in non-computer fields to explore the possibility of cross-disciplinary research. This will especially motivate students in art-related major to explore new ways of expressing art. Lastly, it will contribute to the area of face recognition by using a real-life dataset of over 3,000 portraits. Technologies developed in this study can be applied in a variety of applications. As an example, we show the algorithm's potential contribution to e-business in finding misuse of celebrities' portraits.

Fig. 1. Photos from Brunel photography's exhibition (Chase Jarvis Photography 2018)

2 Literature Review

Face recognition has experienced rapid development in recent years with deep learning infrastructure. With proper training, computer algorithms may perform as well as or even better than human experts in identifying similar faces. These face recognition algorithms provide foundations for our study.

Many research studies use common dataset to test their algorithms. For example, the CMU Multi-PIE Face Database contains more than 750,000 images of 337 people (http://www.multipie.org/). The CAS-PEAL face database contains 99,594 images of 1040 Chinese individuals (595 males and 445 females) (http://www.jdl.ac.cn/peal/index.html). The 10 k US Adult Faces Database is a collection of 10,168 natural face photographs for 2,222 of the faces (http://www.wilmabainbridge.com/facememorability2.html). While these datasets provide great resources for face recognition studies, algorithms are often trained for a particular dataset and the generalizability is unknown. Thus, a new face recognition dataset will contribute to the validity of current algorithms.

Face detection is an essential step before face recognition. A classic method of face detection is Haar Cascaded and AdaBoost algorithms. It was first proposed by Viola and Jones for rapid object detection (Viola et al. 2001). It enables rapid object detection using a boosted cascaded set of simple features and achived an accuracy of above 80% (Viola and Jones 2004). In recent years, CNN-based face detection has been the mainstream approach for object detection and recognition. Among them, AlexNet was a pioneer and won the ImageNet Large Scale Visual Recognition Challenge (Krizhevsky et al. 2012). Zhang et al. (2014) invented a Tasks-Constrained Deep Convolutional Network (TCDCN) to detect bounding boxes of human faces and face landmarks. TCDCN achieved higher detection accuracy than Cascaded CNN and the approach also yielded a significantly lower computational cost. MTCNN is a more recent CNN-based model that combines deep multi-task learning and cascaded networks. Its has achieved performance than TCDCN and Cascaded CNN alone. Besides MTCNN, there are some special face detection algorithms such as Gabor filter (Suri et al. 2011) which performs well under different light effect. But for a generic scenario, CNN-based detections are still the most popular.

Face recognition phase relies on extracting important features on human faces. This is done by performing supervised learning in CNN with an appropriate loss function. The identification of an ideal loss function is critical to the learning performance. One popular loss function is the softmax loss function, which generate rich face features for recognition. Researchers also combine multiple loss functions to achieve better learning performance. For example, Smirnov et al. (2017) combined softmax loss with margin-based loss. The final loss function is $L2S + \lambda\ MB$. (λ is the ratio). By tuning the ratio, one can achieve a higher performance by joint supervision.

There are several challenges in face recognition. Different poses (Logie et al. 1987), lightings (Georghiades et al. 2001), expressions (Sim et al. 2002) and occlusions (such as glasses) will affect face recognition largely. Thus, standardization of images is also a critical step. This step is called alignment (Taigman et al. 2014).

There are few studies look into doppelganger problem. Existing studies only used common dataset. Smirnov et al. (2017) proposed doppelganger mining based on CMU's Multi-PIE database of 337 people faces, a much smaller dataset compare to ours. We are interested in applying face detection and face recognition algorithms in finding doppelgangers on a university campus in China. The technical focus of the study is the integration of available techniques and the specific challenges dealing with this new dataset.

3   Proposed Methodology

In order to find doppelgangers, we propose a four-phase framework that consists of: (1) face detection, (2) face alignment, (3) face recognition and (4) similarity computation as shown in Fig. 2. Face detection component identifies the position of the face. Face alignment component rotates the face to allow cross-image comparison. Face recognition trains a feature extraction model and finally similarity computation is performed. We explain each component in this section.

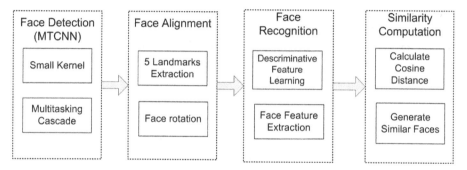

Fig. 2. Framework of Doppelganger mining

3.1   Face Detection (MTCNN) and Alignment

The purpose of face detection is to identify bounding boxes and landmarks from the image of faces. Bonding boxes are imaginary rectangles around an object face. Face landmarks are important points on faces that represents salient regions of the face. Popular landmarks include eyes, eyebrows, nose, mouth, and jawline. Based on our review, we chose to use a Multitasking Convolutional Neural Network (MTCNN) (Zhang et al. 2016). The algorithm is effective and combines the advantage of deep multi-task learning and cascaded network. MTCNN marks five landmarks: two eyes, tip of the nose, and two corners of the mouth.

We used a small convolution kernel (size of 3 * 3 and 2 * 2) in our neural network for several advantages. The small kernel allows us to have more hidden layers and more nonlinear functions, and thus improves the discretion of decision function and reduces the number of parameters to be trained (Simonyan et al. 2014). Because MTCNN contains multiple cascaded networks, number of parameters will greatly slow down training performance. It means the small kernel will allow us to train a fairly complex model with reasonable computing resources.

MTCNN is a sequence of network where the output from the previous network is passed on to the input of the next network. Similar to the AdaBoost algorithm, MTCNN iterates through a sequence of weak classifier to generate a strong classifier and detect five accurate landmarks. Following Zhang's work (2016), we use three networks to generate accurate landmarks: Proposal Network (P-Net), Refine Network (R-Net) and Output Network (O-Net) and the process is illustrated in Fig. 3.

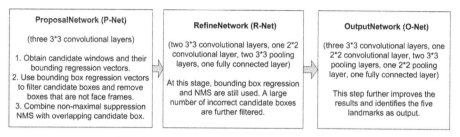

Fig. 3. MTCNN algorithm with three networks

MTCNN identifies three important descriptors for training: face classifier, face founding box, and five landmarks. Face classifier identifies whether the object is a face or not. It can be achieved by a cross entropy loss function shown in formula (1). For each picture i, p_i is the probability of a human face being detected, where y_i^{det} is a background label that is marked in advance from a standard training set.

$$L_i^{det} = -\left(y_i^{det} log(p_i) + \left(1 - y_i^{det}\right)(1 - log(p_i))\right)$$
$$y_i^{det} \in \{0, 1\}$$

(1)

Once an object is classified as a face, face bounding box will be calculated based on Euclidean distance as shown in formula (2). Here, $\hat{y}$ is the result predicted by the network. y is the actual coordinates labeled previously, which is a quaternary:

$$\begin{bmatrix} [x_{upper\ left\ corner\ of\ bounding\ box} & y_{upper\ left\ corner\ of\ bounding\ box} \\ long\ of\ the\ bounding\ box & wide\ of\ bounding\ box \end{bmatrix}$$

$$L_i^{box} = \left|\hat{y}_i^{box} - y_i^{box}\right|_2^2$$
$$y_i^{box} \in R^4$$

(2)

Finally, five key points of the face, or landmarks are positioned as shown in formula (3). Here, $y_i^{landmark}$ represents five landmark points, using a pair of values (x, y). $\hat{y}$ is the predicted face landmarks, and y is the actual marked landmarks.

$$L_i^{landmark} = \left|\hat{y}_i^{landmark} - y_i^{landmark}\right|_2^2$$
$$y_i^{landmark} \in R^{10}$$

(3)

With above three descriptors, the supervised function of final training network is illustrated in formula (4). The goal is to minimize the Euclidean distance between the actual marks and the predicted marks. The result of the three networks is five predicted landmarks.

$$min \sum_{i=1}^{N} \sum_{j \in \{det, box, landmark\}} \alpha_j \beta_i^j l_i^j$$
$$\beta_i^j \in \{0, 1\}$$

P - Net R - Net ($\alpha_{det} = 1, \alpha_{box} = 0.5, \alpha_{landmark} = 0.5$)
O - Net ($\alpha_{det} = 1, \alpha_{box} = 0.5, \alpha_{landmark} = 1$)

(4)

N: number of training samples
α_j: the importance of this loss function
β_i: sample label
Li^j: one of three loss functions mentioned before

3.2 Face Alignment

After the detection of 5 predicted landmarks, pictures need to be rotated and resized for accurate similarity comparison. We first obtain standard landmark positions by aggregating all pictures and generating a group of mean landmarks. This is used as the standard to align all pictures. The five landmarks are then positioned to the closest position to the mean standard landmarks. Figure 4 shows the process of (1) detecting 5 landmarks and (2) face alignment and cropping. The tilted face in the original picture on the left is aligned and cropped and becomes correctitude.

Fig. 4. (. From left to right) Original picture, Face Detection, Face Alignment

3.3 Face Recognition

After detection and alignment, the next step is to train the face recognition model. We chose to train the model with Caffe framework, a deep learning framework developed by Berkeley AI Research (https://caffe.berkeleyvision.org/). Following Wen et al. (2016), we use typical Convolutional Neural Networks with joint supervision of softmax loss and center loss. Face recognition use inter-class and intra-class variations to identify faces from pixels. External factors such as light, covers and expressions may cause the same person to appear differently. Inter-class variation is used to distinguish individuals. Intra-class variation is used to overlook these factors and to identify the same person. We can obtain inter-class dispension by softmax and intra-class compactness by center loss function. Figure 5 illustrates the network structure in Caffe. The convolutional layer is followed by the PReLU (Parametric rectified Linear Unit) (He et al. 2015) as the activation function (Cybenko. G 1989). PReLU is a type of ReLU added parametric rectified (Jarrett et al. 2009). As illustrated in Fig. 5, center loss is introduced in the feature layer output to reach the intra-class aggregation and the inter-class separation.

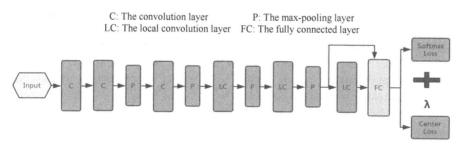

Fig. 5. Face recognition network structure diagram

softmax loss is calculated using formula (5)

$$L_s = -\sum_{i=1}^{m} log \frac{e^{w_{y_i}^T x_i + b_{y_i}}}{\sum_{j=1}^{n} e^{W_j^T x_i + b_j}} \tag{5}$$

X_i - *the ith deep feature in the d-dimensional space. Belongs to the y_i class;*
d - *the dimension of the feature space;*

W -   Fully connected layer parameter matrix. $W = \{d * n\}$. d rows and n columns;
Wj -   the jth column of W;
M -   batch size of mini- batch; N - Number of classes; b - bias

It can be seen that the denominator is all classes and the numerator is a single class.

Formula (6) represents center loss calculation, where c_{yi} is the central feature of category y_i, x_i represents the characteristics before the fully connected layer, and m represents the size of the mini-batch. We want to minimize the distance quadratic sum between sample's features and the center of the features.

$$L_c = \frac{1}{2} \sum_{i=1}^{m} |x_i - c_{y_i}|_2^2 \tag{6}$$

In order to minimize the distance between classes, we use the gradient descent method. Gradient formula is shown in formula (7).

$$\frac{\alpha L_c}{\partial x_i} = x_i - c_{y_i} \tag{7}$$

$$\Delta c_j = \frac{\sum_{i=1}^{m} \delta(y_i = j) \cdot (c_j - x_i)}{1 + \sum_{i=1}^{m} \delta(y_i = j)}$$

Finally, we combine softmax and center loss. We use λ to balance the two loss functions. The larger the value, the greater the intra-class discrimination.

$$L = L_s + \lambda L_c = - \sum_{i=1}^{m} log \frac{e^{w_{y_i}^T x_i + b_{y_i}}}{\sum_{j=1}^{n} e^{W_j^T x_i + b_j}} + \frac{\lambda}{2} \sum_{i=1}^{m} |x_i - c_{y_i}|_2^2 \tag{8}$$

3.4 Similarity Calculation

The final step to find doppelgangers is image similarity calculation. Principal Component Analysis (PCA) is used to reduce feature dimensions (Sirovich et al. 1987). Cosine similarity is a popular similarity calculation that measures the cosine angle between two non-zero vectors of an inner product space (Nguyen et al. 2011). The smaller the angle, the more similar the two vectors are. Thus, a cosine similarity approaching 1 means the two vectors are extremely similar. Formula (9) is the cosine similarity calculation, where F1 and F2 represent extracted feature vectors from face recognition.

$$sim(F1, F2) = cos\theta = \frac{F1 \cdot F2}{|F1| \cdot |F2|} \tag{9}$$

4   Experiments

To evaluate the model, we conducted two experiments, the first experiment on a *labeled* common dataset, the LFW Deep Funneled Images(http://vis-www.cs.umass.edu/lfw/). This experiment is to evaluate our face recognition model, including the detection, alignment and recognition phases. The second experiment is based on an *unlabeled* dataset of over 3,000 students from a school campus. In this experiment, we qualitatively measure the performance of doppelganger mining.

4.1 Experiment 1: Validating Face Recognition Using LFW Dataset

In the first experiment, we used a common dataset from LFW Deep Funneled Images (Huang et al. 2017). The dataset contains a total of 13,233 labeled pictures and is available at http://vis-www.cs.umass.edu/lfw/. The performance is reported using a classification algorithm that identifies multiple pictures of the same person. Notice that the dataset is already labeled with the same person for classification purpose. Although similarity algorithm cannot be directly validated here, the dataset is a good labeled source to validate face recognition algorithm. Furthermore, we examined the influence of face alignment in improving the recognition results. The alignment phase calibrates face position according to the five standard mean landmarks.

Table 1 illustrates the classification results with and without alignment adjustment. Figure 6 illustrates the ROC curve with alignment performed on training dataset and testing dataset. It shows that alignment greatly improve the performance in terms of average precision, AP, EER and TPR (true positive rate). Average precision improved from 78.6% to 96.6% when alignment is properly performed. This experiment validates the performance of our face recognition model with proper alignment algorithm.

Table 1. Validation analysis of alignment performance

	AP (Average Prec.)	EER (equal error rate)	TP rate @0	TP rate @001	TP rate @0001
Without alignment	78.6	67.3	1.07	12.7	6.8
With alignment for training data	82.5	78.8	19.9	31.3	26.5
With alignment for training and testing data	96.8	97.2	71.7	92.3	89.4

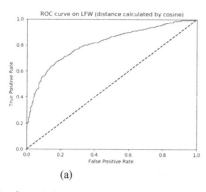
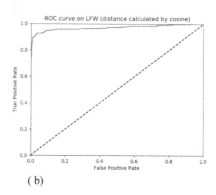

(a)   (b)

Fig. 6. (a) ROC curve with alignment performed for training only; (b) ROC curve with alignment performed for both training and testing

4.2 Experiment 2: Finding Doppelgangers

In experiment 2, we obtained a collection of 3,611 student photos for doppelganger mining. Similar to experiment 1, we performed MTCNN-based face detection, face alignment, and a CNN-based face recognition algorithm with softmax loss and center loss. Finally, cosine similarity was performed on all pair-wise images and similarity score above 0.8 were selected. We calculated a total of over 6.5 million pairs and this threshold gave us a total of 1,218 pairs of ifaces, which accounts for 0.01% of total pair-wised sample as shown in Table 2. We found that five pairs with cosine score higher than 0.9 are actually from the same person, as illustrated in Fig. 7. They submitted photos both during undergraduate and graduate studies. These five pairs were discarded. The fact that the algorithm can successfully identify the same person shows that our face recognition and similarity measure algorithms can be used in other applications with a higher threshold.

Table 2. Similarity Score Distribution and Doppelganger Selection

Total number of images	3,611	
Total number of image pairs	6,517,855	
# of images with cosine score > 0.8	**1,218**	**0.01% of total sample**
# of images with cosine score > 0.9	5	discarded

Fig. 7. Pictures from the same student with cosine similarity score of 0.905

We then manually checked the 1,218 pairs of images and group them into two categories: pairs that are highly alike, or doppelgangers; and pairs that are somewhat alike, not real doppelgangers. The results are concluded to Fig. 8. However, we did not perform a formal tagging to generate a gold standard from the dataset. This manual check was explorative. Figure 9 shows 6 pairs of top-ranked face photos with cosine similarity greater than 0.8 and are judged by human as highly alike pairs. These student pairs are considered true doppelgangers in our study. We plan to organize a photo shooting session with these students to mimic Brunel photography's exhibition.

For the somewhat alike but not real doppelganger group, we further investigated the possible reasons of their high cosine similarity score. We identified the following areas that could cause high cosine similarity while the two faces are not highly alike.

Cosine
Similarity
Score
$\Big\{$
Similarity > 0.9 → **same person**

0.8 < Similarity < 0.9 → highly alike person → **True Doppelgangers**

0.8 < Similarity < 0.9 → somewhat alike person → Alike but **not Doppelgangers**

Fig. 8. Doppelganger identification

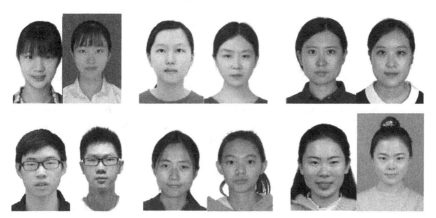

Fig. 9. Doppelgangers on campus: top-ranked face pairs (0.7< cosine similarity <0.85 combination)

(1) The model is sensitive to facial expression. As shown in Fig. 10, the two people smile the same way have a higher chance of being matched with cosine similarity greater than 0.8.

Fig. 10. Matching results affected by facial expressions such as smile styles

Fig. 11. Matching results affected by post-processing such as skin brightening

(2) The model is sensitive to brightened skin, especially with pictures that are post-processed. For example, in Fig. 11, both skin colors were brightened, and facial flaws were blurred. Although they are shown to have high cosine similarity, the two people are less alike. We believe during PCA phase, such skin processing could largely affect the performance of how dimensions are mapped and reduced.

(3) The model is sensitive to facial lines (wrinkles). As shown in Fig. 12, both faces have deep smile lines and slightly raised mouths. These two features become dominant features. This also means, when people age, the recognition algorithm may fail

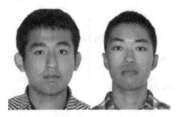

Fig. 12. Matching results affected by facial lines

Fig. 13. Matching results affected by wearing eyeglasses

to recognize them with wrinkles. The sensitivity level can be adjusted during the learning process.

(4) The model is sensitive to accessories on faces. As shown in Fig. 13, two people wearing similar eyeglasses are selected as top-ranked results. The shape and color of glasses are taken into the model as important features in similarity measure.

(5) There is also a dilemma with alignment. We found one face image that has very close landmarks with the mean standard landmarks. As a result, the model identified 12 other face images that are lookalikes. We consider this to be a false positive example.

5 Discussions

Although the focus of this study is on doppelganger mining, our algorithm can be extended to other applications in e-business. However, the threshold for similarity measure needs to be tested. Figure 14 shows two images of the same person Gigi Hadid. Picture on the left is a picture used in a store in Taobao platform. Picture on the right is a runway picture of Hadid. The similarity score here is 0.715. Our testing results on LFW image database also shows that on average, pictures of the same persons' similarity

Fig. 14. Catwalk picture similarity calculation (with cosine similarity = 0.715)

score are 0.5–0.8. Our experiment 2 is based on students' ID photos and the similarity threshold should be higher.

6 Conclusions

This study aims to perform doppelganger mining by using MTCNN-based face detection and CNN-based face recognition algorithm. MTCNN is an effective method for face detection. Both softmax loss and center loss functions are used in face recognition to identify the best face features. The model is trained on Caffe framework. Using a dataset of over 3,000 face images, we are able to identify pairs of doppelgangers that do look alike each other. We also found that the model is sensitive to facial expression, skin brightness, skin lines and accessories on face.

In the future, we would like to exhibit the results of this study in the form of pairs of pictures, similar to Brunel Photography's Exhibition. This will inspire students of non-CS major to explore deep learning and image recognition technology. Furthermore, it will encourage collaboration across-disciplines on campus. In terms of e-business application, the similarity comparison algorithm shows great potential in detecting violations of portrait right. We also plan to further improve our model to address the current four problems we observe, and also apply the model to more application domains.

References

Chase Jarvis Photography (2018). https://www.chasejarvis.com/photos/

Cybenko, G.: Approximation by superpositions of a sigmoidal function. Math. Control, Signals Syst. 2(4), 303–314 (1989). https://doi.org/10.1007/BF02134016

Huang, G.B., Ramesh, M., Berg, T., Learned-Miller, E.: Labeled faces in the wild: a database for studying face recognition in unconstrained environments. University of Massachusetts, Amherst, Technical Report 07–49, October 2007

Georghiades, A.S., Belhumeur, P.N., Kriegman, D.J.: From few to many: illumination cone models for face recognition under variable lighting and pose. IEEE Trans. Pattern Anal. Mach. Intell. 6, 643–660 (2001). https://doi.org/10.1109/34.927464

He, K., Zhang, X., Ren, S., Sun, J.: Delving deep into rectifiers: surpassing human-level performance on imagenet classification. In: Proceedings of the IEEE International Conference on Computer Vision, pp. 1026–1034 (2015)

Jarrett, K., Kavukcuoglu, K., LeCun, Y.: What is the best multi-stage architecture for object recognition?. In: 2009 IEEE 12th International Conference on Computer Vision, pp. 2146–2153. IEEE, September 2009. https://doi.org/10.1109/iccv.2009.5459469

Krizhevsky, A., Sutskever, I., Hinton, G.E.: Imagenet classification with deep convolutional neural networks. In: Advances in Neural Information Processing Systems, pp. 1097–1105 (2012)

Logie, R.H., Baddeley, A.D., Woodhead, M.M.: Face recognition, pose and ecological validity. Appl. Cogn. Psychol. 1(1), 53–69 (1987)

Nguyen, H.V., Bai, L.: Cosine similarity metric learning for face verification. In: Kimmel, R., Klette, R., Sugimoto, A. (eds.) ACCV 2010. LNCS, vol. 6493, pp. 709–720. Springer, Heidelberg (2011). https://doi.org/10.1007/978-3-642-19309-5_55

Simonyan, K., Zisserman, A.: Very deep convolutional networks for large-scale image recognition. arXiv preprint arXiv:1409.1556 (2014)

Sirovich, L., Kirby, M.: Low-dimensional procedure for the characterization of human faces. Josa a **4**(3), 519–524 (1987). https://doi.org/10.1364/JOSAA.4.000519

Smirnov, E., Melnikov, A., Novoselov, S., Luckyanets, E., Lavrentyeva, G.: Doppelganger mining for face representation learning. In: Proceedings of the IEEE International Conference on Computer Vision, pp. 1916–1923 (2017)

Suri, P.K., Walia, E., Verma, E.A.: Novel face detection using Gabor filter bank with variable threshold. In: 2011 IEEE 3rd International Conference on Communication Software and Networks, pp. 715–720. IEEE, May 2011. https://doi.org/10.1109/iccsn.2011.6014992

Sim, T., Baker, S., Bsat, M.: The CMU pose, illumination, and expression (PIE) database. In: Proceedings of Fifth IEEE International Conference on Automatic Face Gesture Recognition, pp. 53–58. IEEE, May 2002. https://doi.org/10.1109/afgr.2002.1004130

Taigman, Y., Yang, M., Ranzato, M.A., Wolf, L.: Deepface: x. In: Proceedings of the IEEE Conference on Computer Vision and Pattern Recognition, pp. 1701–1708 (2014)

Time (2018). https://time.com/5435683/artificial-intelligence-painting-christies/

Viola, P., Jones, M.: Rapid object detection using a boosted cascade of simple features. In: CVPR, vol. 1, pp. 511–518, 3 (2001). https://doi.org/10.1109/cvpr.2001.990517

Viola, P., Jones, M.J.: Robust real-time face detection. Int. J. Comput. Vis. **57**(2), 137–154 (2004). https://doi.org/10.1023/B:VISI.0000013087.49260.fb

Wen, Y., Zhang, K., Li, Z., Qiao, Yu.: A discriminative feature learning approach for deep face recognition. In: Leibe, B., Matas, J., Sebe, N., Welling, M. (eds.) ECCV 2016. LNCS, vol. 9911, pp. 499–515. Springer, Cham (2016). https://doi.org/10.1007/978-3-319-46478-7_31

Zhang, Z., Luo, P., Loy, C.C., Tang, X.: Facial landmark detection by deep multi-task learning. In: Fleet, D., Pajdla, T., Schiele, B., Tuytelaars, T. (eds.) ECCV 2014. LNCS, vol. 8694, pp. 94–108. Springer, Cham (2014). https://doi.org/10.1007/978-3-319-10599-4_7

Zhang, K., Zhang, Z., Li, Z., Qiao, Y.: Joint face detection and alignment using multitask cascaded convolutional networks. IEEE Signal Process. Lett. **23**(10), 1499–1503 (2016). https://doi.org/10.1109/LSP.2016.2603342

Time Series Analysis of Open Source Projects Popularity

Shahab Bayati[1]([⊠]) and Marzieh Heidary[2]

[1] The University of Auckland, Auckland, New Zealand
s.bayati@auckland.ac.nz
[2] Spark New Zealand, Auckland, New Zealand
m.heidary@gmail.com

Abstract. Open source software (OSS) community relies on volunteers and developers contributions for its survival. However, only a few projects reach success and popularity in open source community. Then, it is important to know the success factors of OSS projects. In this paper, we have applied time series clustering on open source projects hosted on a social coding platform to understand the main effective attributes of an OSS project on its popularity trends. We have applied exploratory data analysis on each cluster to see the effect of projects' performance and attributes on projects' reputation inside the OSS community. Finally, we have applied machine learning techniques to predict the popularity trend of OSS projects. Having access to the social coding data expands our view on project popularity on both social and technical factors. Results of this empirical study can help project owners and members to manage and promote the project reputation.

Keywords: Social coding · Mining software repositories · Time series clustering · Open source software (OSS) · Prediction

1 Introduction

Open source software (OSS) and open innovations are getting popular subjects in information systems and software engineering literature. The great success of Android, Linux, R, Firefox, and Apache projects is just one side of open source projects. On the other side, millions of open source projects are failed and stopped. Various studies in the open source literature focused on the success factors and different key elements are found (Grewal et al. 2006; Subramaniam et al. 2009; Wu and Goh 2009). Along with all of the factors studied in the literature, developers' attraction and developers' sustainable commitment are mentioned as an important factor for project success (Schilling et al. 2012).

Open source development is not just a technical process and as many volunteers across the globe are contributing and interacting to OSS community it is considered as a socio-technical process (Dabbish et al. 2013; Tsay et al. 2014). As a reason, social coding platforms emerge. The largest online social coding platform is GitHub with more

© Springer Nature Switzerland AG 2020
K. R. Lang et al. (Eds.): WeB 2019, LNBIP 403, pp. 77–86, 2020.
https://doi.org/10.1007/978-3-030-67781-7_8

than 100 million projects 24 millions of registered users. In this study, we collected GitHub's data to form a time series of OSS projects popularity trend among developers. We used this data to find the most effective project attributes on project popularity and reputation. We have applied mining software repositories (MSR) techniques to GitHub data to find the most important projects attributes on popularity trends. Then the main research questions of this study are "RQ1: Which OSS project socio-technical attributes are affecting the project popularity trend among OSS developers?" and "How to predict the popularity trend of OSS project using a snapshot data?"

This study contributes to the field of empirical software engineering and OSS in terms of using a longitudinal dataset of projects in a social coding environment to see the most effective socio-technical activities on project popularity. Lack of longitudinal studies in GitHub mining literature is mentioned in (Cosentino et al. 2017). Also, we have focused on project success through its reputation in the community of OSS developers compare to previous studies focusing on download counts, page view and subscribers (Grewal et al. 2006; Subramaniam et al. 2009; Wu and Goh 2009) as a popularity measure. We have considered the developers level of interest.

This paper is structured in a way that after introduction, related works are discussed. Data collection and pre-processing are discussed later. Then, the main data analysis and interpretation are presented. Finally, we have concluded our work and future research trends are discussed.

2 Related Works

In this section, we briefly review the related works to this study. We focus on mining software repositories literature and then review on OSS project success and popularity studies. This helps us to define our research contribution in relation to OSS literature.

The importance of observing developers socio-technical activities in the OSS community and in social coding platform is highlighted in (Dabbish et al. 2013; Tsay et al. 2014). They have mentioned the projects and developers transparent activities in public OSS repositories make decision-making process for practitioners and organizers easier. The effect of social tie strength in OSS community on open source project success is investigated in (Grewal et al. 2006). In addition, developers commitment (Subramaniam et al. 2009), project age (Lee et al. 2009) project complexity (Midha and Palvia 2012), organizational sponsorship, and license type (Stewart et al. 2006) are revealed as success factor determinants. Most of these studies measured success with external factors such as commitment level, download count (Grewal et al. 2006; Subramaniam et al. 2009; Wu and Goh 2009). Project reputation across the OSS community is not used in these studies. This study analyzes the project's popularity among OSS developers by using "Watching" event in GitHub.

Popular developers are studied in the OSS context in (Blincoe et al. 2016). Following popular developers as a source of trust helps developers to find trustworthy and useful resources. Also, they are studied for the purpose of expert recommendation (Schall 2014). Few studies investigated popular projects on GitHub. An exploratory analysis of top Python projects reveals the importance of documentation in OSS popularity (Weber and Luo 2014). Also (Jarczyk et al. 2014) look for popular projects on GitHub by applying statistical analysis on a snapshot of GitHub projects events. (Borges et al. 2016) also tried to explore popular projects in GitHub. Longitudinal data is not used in literature to find the most effective factors on GitHub projects popularity.

3 Data Processing

This study analyzes the popularity of open source projects over the time, we need to extract related data from GitHub. GitHub provides a RESTful API for accessing public data. GhTorrent (Gousios and Spinellis 2012) is a data dump of GitHub which mirrored most of the events and activities in GitHub repositories by using GitHub API. We used GhTorrent data dump in this study. It contains data and Metadata about GitHub repositories for more than 8 years.

We applied data pre-processing on this data dump as many of GitHub projects are not real software development projects. A big range of projects are deleted from GitHub server or they are not active anymore (Kalliamvakou et al. 2014). For the purpose of this study, we selected data of GitHub projects which are created in different months in 2012 and we focused on projects with more than 10 developers (Yamashita et al. 2016) as core members. These constraints are applied to have a more realistic set of open source projects where we can see the role of each individual in the development process. In this study, we have collected projects written in the most popular programming languages in GitHub community. These programming languages include Java, JavaScript, C++, Python, PhP, and C#. Table 1 shows the numbers of repositories for each of these languages in our prepared dataset. Also, all of the collected projects are the original root projects and we ignored all the forked and mirrored projects. Outliers are removed and finally, we reached the 343 projects.

Table 1. The number of projects developed in each programming language in our data set.

C#	C++	Java	JS	PhP	Python	Total
10	22	76	104	38	93	343

For the purpose of time-series analysis, we collected the monthly data of project watchers as a proxy of popularity among developers for a period of two years (24 Months) after creation date. In addition, we collected projects' attributes after a year of creation to see which attributes play an important role in projects popularity. In this dataset, we captured the number of watchers a project gained for each month. We considered "Watching" event in GitHub as a factor of popularity among the community members. In GitHub community, members may watch a project when they are interested in that project and want to get updated information about the events of the watched project. By collecting this data we built the foundation for our time series analysis. Then by the supporting of GhTorrent data dump, we calculated the number of submitted commits to each project after a year of creation. A number of commits are used in many studies as a proxy for OSS project activity (Subramaniam et al. 2009; Wu and Goh 2009). We also added a number of issues and bug reports registered for each project. A number of committers who contributed to the project development process is considered in our dataset. Committers can be project members or volunteers from the community. Furthermore, we used pull request (Yu et al. 2015) information. Also, as GitHub is a social coding platform we have added all the comments applied to coding activities.

To gain more data about what was happened in the project we calculated the number of time a project is forked by others inside a community for further contribution or developing a divergence project (Jiang et al. 2017). For each project, we checked for the owner type. In GitHub, a project owner can be an individual user or an organization. Furthermore, for each project, the main programming language is collected. We have converted the collected data in the format of CSV for data analysis process which is discussed in the next section.

4   Data Analysis

Our main strategy for data analysis is applying the unsupervised technique on the time series data collected for project popularity based on a number of watchers. We applied Dynamic Time Wrapping (DTW) clustering for clustering time series based on their shape similarities and other optimization and transformation technique (Oates et al. 1999). We use the "dtwclust" R package for this purpose. We have created 4 clusters. Figure 1 shows the clustering result for the accumulative number of watchers for each project.

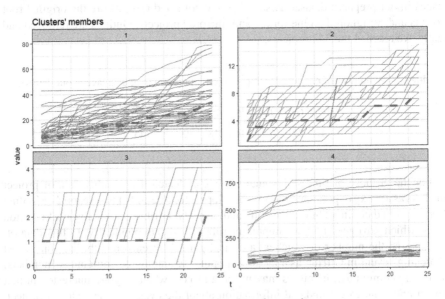

Fig. 1. Time series clustering result

As the clustering result shows in cluster 3 we have just project with lowest number of watchers and the centroid trend is almost flat in this cluster. Cluster 4 has the most number of watchers. Some projects in this cluster reach to more than 500 watchers. The detail of cluster analysis and witch factors has the most important role in defining a project position in each cluster is discussed later. Figure 2 demonstrates the cluster centroids trends. Cluster 4 starts with 40 watchers and ended with 120 after two years.

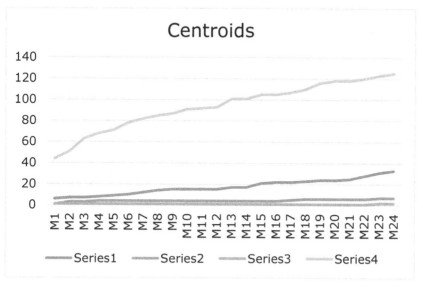

Fig. 2. Cluster centroids

To see the distribution of watchers on each cluster, boxplots are visualized in Fig. 3. The red line represents the trend of the median of watchers from one month to another month. Vertical boxes show the distribution of watchers for each month.

4.1 Effective Factors Analysis

To understand the main factors correlated with the allocated clusters of popularity in OSS projects, we have applied Random Forest on our collected dataset. We have chosen clusters as a dependent variable and other projects characteristics as independent variables. Figure 4 illustrates the result of Random Forest analysis. Random Forest is an ensemble machine learning technique which can be used to determine feature importance in prediction problems.

Figure 4 illustrates the main important factors lead to different popularity trend clusters. This chart is divided into two parts. On the left side mean decrease accuracy shows how worse classification is on the absence of each factor and it reveals the importance of that factor for decision making. On the right side mean decrease Gini calculate the decrease in purity of results on decision trees nodes without considering the attributes. Based on the result of random forest analysis, the most important factor defines the clusters is the time's project forked. Comments on issues, Pull requests and commit counts are the other important factors on both charts. Owner type is the least important factor.

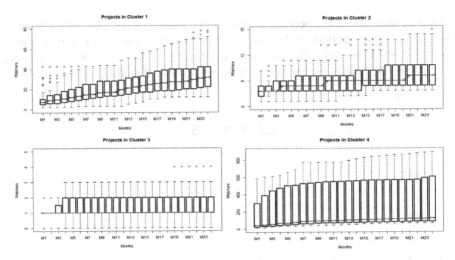

Fig. 3. Changes in project watchers distribution over time in each cluster

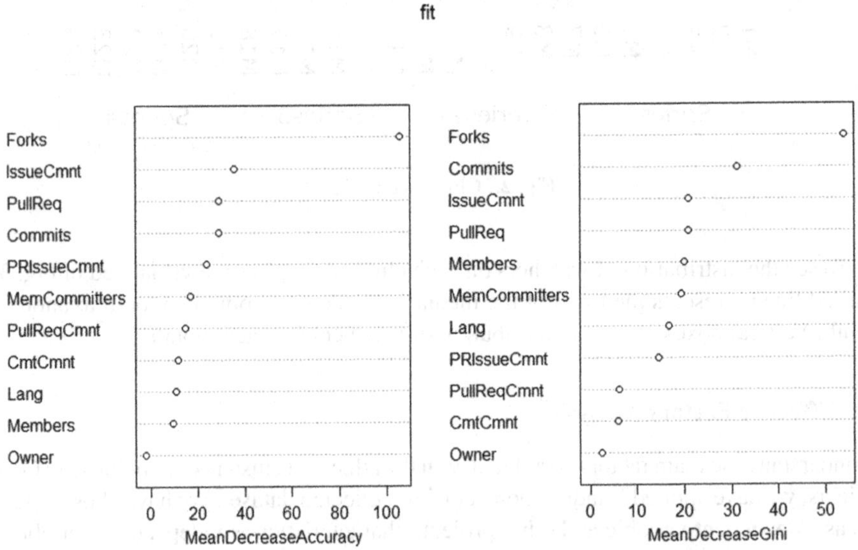

fit

Fig. 4. Feature importance results from random forest analysis

To have a brief understanding of each cluster and compare it with other clusters regarding the projects' attributes we have provided the boxplots diagrams to visually see the distribution of projects factors in each cluster. Figure 5 visually compares each time-series cluster of watchers with other project attributes. It can be revealed that cluster 4 and after that cluster 1 have the highest values on each cluster.

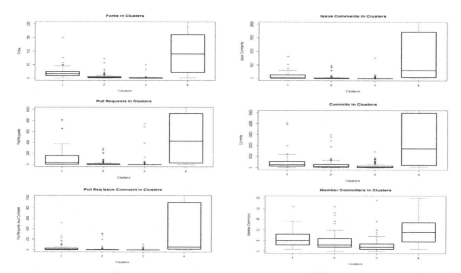

Fig. 5. Comparing clusters via boxplot diagram (Forks, Issue Comments, Pull Requests, Commits, Pull Request Issue Comments, and Member Committers)

Also, to compare clusters based on the main programming language of the projects we have provided Table 2. Table 2 shows how different programming languages are distributed over clusters. Table 3 shows different owner types are distributed over clusters.

Table 2. Popularity clusters and programming languages

Cluster	C#	C++	Java	JS	PhP	Python	Total
1	2	1	11	24	4	8	50
2	3	11	28	35	7	29	113
3	5	8	35	37	22	53	160
4	0	2	2	8	5	3	20
Sum	10	22	76	104	38	93	343

Totally 30% of the projects in our dataset belongs to JavaScript community almost 40% of cluster 4 and 50% of cluster1 projects are based on JavaScript. 11% of projects are written in PhP but 25% of the most popular ones are in PhP. Python has the largest number of unpopular projects in cluster 3. This exploratory analysis shows that programming language can affect popularity.

As we have chosen projects with more than 10 core members in our analysis, more than 90% of projects are organization founded projects. This will be different if we include smaller projects. However, it shows that owner type is not effective in popularity. But we can argue owner type has an effect on project size which requires further investigations.

Table 3. Popularity clusters and owner types

Cluster	Organization	Individual user	Total
1	45 (90%)	5 (10%)	50
2	103 (91%)	10 (9%)	113
3	148 (92%)	12 (8%)	160
4	19 (95%)	1 (5%)	20
Sum	315(92%)	28 (8%)	343

4.2 Prediction

In this section, we aimed to predict the popularity trend of projects with the help of machine learning techniques. We have used Weka as an open source machine learning platform to predict each project cluster allocated to the related OSS project. We have used the project attributes after a year of creation as independent variables. The dependent variable is associated cluster of popularity (watchers) trend to each OSS project. For the supervised classification, we have tested multiple techniques. Naïve Bayes, C4.5, Multi-Layer Perceptron Neural Network (MLP-NN), Multinomial Logistic Regression (MLR), Linear Logistic Regression (LLR), Support Vector Machine (SVM), Ada Boost and Baseline (ZeroR) classifiers are applied. To evaluate the performance of prediction we have used 10-Fold cross-validation technique.

Table 4. Comparison of different classification approaches for popularity trend analysis

Technique	CCIR	TP	FP	Precision	Recall	F-M	ROC
Naïve Bayes	54.81	0.548	0.345	0.537	0.548	0.445	0.683
C4.5	**66.18**	**0.662**	**0.191**	**0.657**	**0.662**	**0.654**	0.732
MLP-NN	46.64	0.466	0.337	0.423	0.466	0.433	0.638
MLR	60.34	0.603	0.248	0.583	0.603	0.583	0.765
LLR	60.64	0.606	0.266	0.583	0.606	0.570	**0.777**
SVM	46.64	0.466	0.432	0.354	0.466	0.363	0.518
AdaBoost	58.60	0.586	0.276	0.359	0.586	0.446	0.665
ZeroR	46.64	0.466	0.466	0.218	0.466	0.297	0.494

To analyze the performance of each prediction technique we have used the weighted average of accuracy metrics for each classification technique. Correctly Classified Instances Rate (CCIR), True positive rate (TP), False Positive Rate (FP), Precision, Recall, F-Measure (F-M), ROC Curve Area (ROC) are the main accuracy metrics used in this study.

Table 4 compares the result of different classification techniques. The bolded values represent the highest and most accurate values among all others. Based on the outcome of the classification analysis, the most accurate prediction technique is C4.5. Highest value of precision and recall among all other techniques make this machine learning technique the most appropriate one to predict the right trend for OSS project popularity.

5 Conclusion, Limitations and Future Works

In this study, we focused on the open source projects reputation among GitHub members by using a time driven dataset of projects watchers. Our analysis of four different clusters of popularity trends shows that technical contribution to the projects through forking and committing are the most effective factors in setting a project on a specific cluster. Social contribution through issues discussion is on the second d level of priority. Our study confirms the finding of previous studies regarding the correlation of forks and watchers. We have applied multiple machine learning techniques to find the most suitable approach to predict the popularity trend among OSS projects. C4.5 shows the best performance across all other techniques.

In this study, we have only focused on major programming languages on GitHub and others are ignored. To have a more generalized view, we can include more programming languages. In this study we only have projects with minimum of 10 core members and the results can be evaluated with the cases of smaller projects. More items about projects can be evaluated in future, such as project domain and context. Project prior to 2012 and after that also can be involved.

References

Blincoe, K., Sheoran, J., Goggins, S., Petakovic, E., Damian, D.: Understanding the popular users: following, affiliation influence and leadership on Github. Inf. Softw. Technol. **70**, 30–39 (2016)

Borges, H., Hora, A., Valente, M.T.: Understanding the factors that impact the popularity of Github repositories. arXiv preprint arXiv:1606.04984 (2016)

Cosentino, V., Izquierdo, J.L.C., Cabot, J.: A systematic mapping study of software development with Github. IEEE Access **5**, 7173–7192 (2017)

Dabbish, L., Stuart, C., Tsay, J., Herbsleb, J.: Leveraging transparency. IEEE Softw. **30**(1), 37–43 (2013)

Gousios, G., Spinellis, D.: Ghtorrent: Github's data from a firehose. In: 2012 9th IEEE Working Conference on Mining Software Repositories (MSR), pp. 12–21. IEEE (2012)

Grewal, R., Lilien, G.L., Mallapragada, G.: Location, location, location: how network embeddedness affects project success in open source systems. Manage. Sci. **52**(7), 1043–1056 (2006)

Jarczyk, O., Gruszka, B., Jaroszewicz, S., Bukowski, L., Wierzbicki, A.: GitHub projects. quality analysis of open-source software. In: Aiello, L.M., McFarland, D. (eds.) SocInfo 2014. LNCS, vol. 8851, pp. 80–94. Springer, Cham (2014). https://doi.org/10.1007/978-3-319-13734-6_6

Jiang, J., Lo, D., He, J., Xia, X., Kochhar, P.S., Zhang, L.: Why and how developers fork what from whom in Github. Empirical Softw. Eng. **22**(1), 547–578 (2017)

Kalliamvakou, E., Gousios, G., Blincoe, K., Singer, L., German, D.M., Damian, D.: The promises and perils of mining Github. In: Proceedings of the 11th Working Conference on Mining Software Repositories, pp. 92–101. ACM (2014)

Lee, S.-Y.T., Kim, H.-W., Gupta, S.: Measuring open source software success. Omega **37**(2), 426–438 (2009)

Midha, V., Palvia, P.: Factors affecting the success of open source software. J. Syst. Softw. **85**(4), 895–905 (2012)

Oates, T., Firoiu, L., Cohen, P.R.: Clustering time series with hidden markov models and dynamic time warping. In: Proceedings of the IJCAI-99 Workshop on Neural, Symbolic and Reinforcement Learning Methods for Sequence Learning, Sweden Stockholm, pp. 17–21 (1999)

Schall, D.: Who to follow recommendation in large-scale online development communities. Inf. Softw. Technol. **56**(12), 1543–1555 (2014)

Schilling, A., Laumer, S., Weitzel, T.: Who will remain? An evaluation of actual person-job and person-team fit to predict developer retention in floss projects. In: 2012 45th Hawaii International Conference on System Science (HICSS), pp. 3446–3455. IEEE (2012)

Stewart, K.J., Ammeter, A.P., Maruping, L.M.: Impacts of license choice and organizational sponsorship on user interest and development activity in open source software projects. Inf. Syst. Res. **17**(2), 126–144 (2006)

Subramaniam, C., Sen, R., Nelson, M.L.: Determinants of open source software project success: a longitudinal study. Decis. Support Syst. **46**(2), 576–585 (2009)

Tsay, J., Dabbish, L., Herbsleb, J.: Influence of social and technical factors for evaluating contribution in Github. In: Proceedings of the 36th International Conference on Software Engineering, pp. 356–366. ACM (2014)

Weber, S., Luo, J.: What makes an open source code popular on Git Hub? In: 2014 IEEE International Conference on Data Mining Workshop, pp. 851–855. IEEE (2014)

Wu, J., Goh, K.Y.: Evaluating longitudinal success of open source software projects: a social network perspective. In: 42nd Hawaii International Conference on System Sciences, HICSS'2009, pp. 1–10. IEEE (2009)

Yamashita, K., Kamei, Y., McIntosh, S., Hassan, A.E., Ubayashi, N.: Magnet or Sticky? measuring project characteristics from the perspective of developer attraction and retention. J. Inf. Process. **24**(2), 339–348 (2016)

Yu, Y., Wang, H., Filkov, V., Devanbu, P., Vasilescu, B.: Wait for it: determinants of pull request evaluation latency on Github. In: 2015 IEEE/ACM 12th Working Conference on Mining Software Repositories (MSR), pp. 367–371. IEEE (2015)

Digital Platforms and Social Media

Social Media or Website? Research on Online Advertising Type Based on Evolutionary Game

Xiang He, Li Li$^{(\boxtimes)}$, Hua Zhang, and Xingzhen Zhu

Nanjing University of Science and Technology, Nanjing, China
lily691111@126.com

Abstract. Choosing an online advertising channel to maximize the benefit of advertisement has become a main topic for advertisers. This paper constructs a model to describe an evolution process, which includes attention cost and persuasive knowledge of consumers. The results show that attention cost paid by consumers and the persuasive knowledge activated by advertisements may have an important impact on the evolutionary results. Specifically speaking, the model will eventually restrain to social media channel because of the higher attention cost paid by consumers and lower persuasive knowledge activated by advertisement; on the contrary, the model will eventually restrain to website channel because of the lower attention cost paid by consumers and higher persuasive knowledge activated by advertising. Furthermore, the model also suggests that the browsing probability of channel would adjust the evolutionary results. The reliability of conclusions is further verified by an example.

Keywords: Social media advertising · Website advertising · Online display advertising · Evolutionary game

1 Introduction

Online display advertising, which includes banner ads, video ads and all nontext-based ads on web pages are exposed through two channels: traditional website and social media. Recently, social media advertising has become much more significant with the rapid development of social media. It is evident that before 31st March 2018, Facebook, as one of the most popular social medias has benefited about \$118 billion during the first quarter in 2018 [1]. In the environment of multi-channels, a primary concern for advertisers choosing a channel has been how to most effectively balance "reject" and "accept". These two behaviors are concerned with the expression style of advertisings: If advertisers choose website channel, its outstanding form different with surroundings could compose consumers quickly recognize the advertising, while consumers would be dissatisfied with the unveiled advertising content; if advertisers choose social media channel, it is not easy for consumers to avoid advertising information because of the same form with other content, while it is difficult for users to understand the intention of the advertising. Despite the advancement of advertising channels, the challenge of choosing a proper channel to balance "reject" and "accept" remains substantial.

© Springer Nature Switzerland AG 2020
K. R. Lang et al. (Eds.): WeB 2019, LNBIP 403, pp. 89–102, 2020.
https://doi.org/10.1007/978-3-030-67781-7_9

Two common measures of channel performance are purchase probability and conversion rate. For example, Breuer and brettel compared website channels (including coupon advertising, banner advertising and price comparison advertising) with search engine channels based on purchase probability, and found that website channels were weaker than search engine channels [2]; Ghose and Todri proposed that the longer advertising exposure time was, the higher purchase probability was [3]; Manchanda et al. study the effects of banner advertising exposure on the probability of repeated purchase using a survival model [4]; Moe and Fader propose a model of accumulative effects of website visits to investigate their effects on purchase conversion [5]. Although above studies could evaluate the effect of channels with online data directly, while there are still some notable gaps. These researches only focus on the last click behaviors before purchasing and pay less attention on the dynamic behavior of how consumers "avoid" and "accept" the advertising, as well as the dynamic channel decision about advertisers [6].

This study aims to bridge this gap by developing an innovative modeling approach that captures the dynamic behaviors of consumers and advertisers. To properly characterize the dynamics of consumers' online behaviors, the model needs the following two parameters. First, different advertising channels have their distinct natures and therefore differ greatly in activating consumers' persuasive knowledge. Based on the persuasive knowledge theory, website channel compose consumers exposed to direct-expression website advertising, which is easy to activate users' persuasive knowledge; On the contrary, social media channel integrates persuasive information into user's social circle, which means users could receive advertising information and social activities at the same time. Since there is no clear boundary among advertising, entertainment and information, it is not easy for social media channel to activate users' persuasive knowledge [6, 7]. Second, consumers' attention cost varies from individual to individual, which could also be affected by different advertising channels. Based on limited attention theory, consumers would not pay more attention on surfing or interacting with the advertising since its short length and simple content, while social media advertising evokes a certain degree of activities with the consumers (such as commenting, praising, etc.) [8], thereby social media attract more attention rather than just passively exposing consumers to website advertising.

Therefore, incorporating the persuasive knowledge and attention contribution activated by different channels in the model is crucial in accurately evaluating the channels' effects. In this paper, we develop an evolutionary game model, which captures different levels of persuasive knowledge and attention contribution activated by website channel and social media channel. The model provides marketing managers and advertises with an innovative perspective to make channel strategies. Moreover, the results from our analysis are helpful to understand the mechanism of persuasive knowledge and limited attention in the selection of online advertising channels, have filled the gaps about advertising channel decisions based on online data empirical data only.

2 Literature Review

In an environment of online advertising, the most researches about online advertising strategies mainly focus on single channel, which include the following three subjects: 1) Channel auctions: Zhu and Wilbur [9] analyzes the hybrid advertising auction, in which each advertiser must choose whether to use CPC bidding or CPM bidding and consider how channels might offer advertisers more efficient mechanisms within the class of hybrid auctions; Arnosti et al. [10] compares the performance of different auction designs when advertiser valuations are positively correlated and advertisers are differentiated in their abilities to estimate the values of single display. 2) Boundary of channel effect: By modeling audience uncertainty, forecast errors, and the ad server's execution of the plan, Turner [11] derives the sufficient conditions about maximize the boundary of channel effect; Zhang et al. [12] studied the boundary of channel effect in different competition conditions based on the content preference and product preference, they concluded that when market competition is moderate, the advertisers would expand the consumers boundary; when competition is high, the advertisers would narrow the consumers boundary. 3) Factors of channels effect: Montgomery et al. [13] develop a Markov model in a given webpage view, they consider that the effect of the channel is affected by the type of the last webpage view; Park and Fader [14] apply the Sarmanov family of bivariate distributions to model the dependence of website visit durations to the effect of the channel. These models provide a new idea about the channel strategy by considering various channel characteristics comprehensively. While existent studies on channel strategy focus solely on single channel, rather than multi-channels, which motivates us to fill this gap.

Online advertising channel strategy is not a one-turn decision; it is a stable result by the evolution of the market in several turns. Most of the existing literature apply evolutionary game model in addressing the strategy choice problems. For example, in the public management field, Li et al. [15] apply the evolutionary game model to analyze the government's strategy choice mechanism in technology, policy, market and other aspects to promote the project subsidy and supervision in China; Liu et al. [16] used evolutionary game theory to demonstrate the initial willingness among carbon emission participants, environmental protection departments and carbon emission advertisers, as well as the impact of policies of environmental protection departments on the evolution of behavior strategies of participants; Wu et al. [17] construct a evolutionary game model among government, universities and advertisers, analyze the strategy choices of the three parties in the process of collaborative innovation.

Thus, evolutionary game is a prevalent approach in strategy selection. Ou [18] first introduced evolutionary game model into online advertising field by constructing a model between consumers and advertisers, consumers and consumers, determined the strategy choice of high browsing or low browsing channels. However, there are still some limitations in this study if the channels are all belong to high browsing channels (such as website channel and social media channel), how advertisers would make appropriate channel strategy is not reflected in this paper. Thus, except the browsing probability, other factors caused by different channels also should be considered. For this reason, a model integrating the persuasive knowledge and attention contribution activated by different channels is especially desirable.

3 The Model

This paper considers a dynamic market, that is, the behaviors of decisions between consumers and advertisers are in the dynamic process with continuous selection and imitation. In this market, an advertiser produces products with quality q and price p. Without considering other sales activities such as discounts, the advertiser has two choices of online display advertising channel strategies: (1) investing website channels such as banners advertising or pop-up advertising (hereinafter referred to website advertisings); (2) investing social media advertisings such as Facebook advertisings (hereinafter referred to social media advertisings). Consumers could get product information through the advertising channel only, thus they also have two different behavior strategies: (1) browsing website advertisings; (2) browsing social media advertisings. Study assumes that the probabilities of the advertiser invests website advertisings or social media advertisings are Y and $1 - Y, (0 \leq Y \leq 1)$, the probabilities of consumers browsing or not browsing are X and $1 - X, (0 \leq X \leq 1)$.

Based on the following strategy, we firstly define q as the initial attention cost, since initial attention cost is related about product's quality. So the attention cost of consumers who browsing website advertisings or social media advertisings are represented with αq and βq, α and β are coefficients of attention cost. From the perspective of attention distribution, social media attract more attention rather than just passively exposing consumers to website advertising, so it assumed that $\alpha q < \beta q$. Moreover, suppose that each consumer with initial persuasive knowledge w, k_1 and k_2 are the coefficients of persuasive knowledge activated by advertising. When people are browsing website advertising, persuasive knowledge would be activated with $k_1 w$, when they are browsing social media advertising, persuasive knowledge would be activated with $k_2 w$. Based on prior study [19], this study may assume that $k_1 w > k_2 w$.

Based on the theory of declining marginal benefits, advertisers would pay $\frac{1}{2}(\alpha q)^2$ for website advertising, $\frac{1}{2}(\beta q)^2$ for social media advertising. Since $\alpha q < \beta q$, therefore $\frac{1}{2}(\alpha q)^2 < \frac{1}{2}(\beta q)^2$. In addition, associated with the classic advertising investment model - reputational model [20], it could be seen that if advertiser needs to pay $\frac{1}{2}q^2$ on advertising, the benefit is going to be $p + \frac{\sqrt{2}}{2}\lambda q$, in detail, λ refers to the browsing probability of the channel. To conclude, the benefit of website advertising and social media advertising are represented with $p + \frac{\sqrt{2}}{2}\lambda_1\alpha q$ and $p + \frac{\sqrt{2}}{2}\lambda_2\beta q$.

In summary, the benefit matrix of consumers and advertisers is shown in Table 1.

Table 1. Consumers and advertisers payoff matrix

	Probability	Website advertising A_1	Social media advertising A_2
		Y	$1 - Y$
Browse (B_1)	X	$(q + wk_1 - \alpha q,$ $p + \frac{\sqrt{2}}{2}\lambda_1\alpha q - \frac{1}{2}(\alpha q)^2)$	$(q + wk_2 - \beta q,$ $p + \frac{\sqrt{2}}{2}\lambda_2\beta q - \frac{1}{2}(\beta q)^2)$
Do not browse (B_2)	$1 - X$	$\left(0, -\frac{1}{2}(\alpha q)^2\right)$	$\left(0, -\frac{1}{2}(\beta q)^2\right)$

4 Equilibrium Analyses

According to Table 1, the expected benefit of two strategies that consumers choose to 'browse' and 'do not browse' as well as the average benefit of the consumer groups are:

$$U_{B_1} = Y(q + wk_1 - \alpha q) + (1 - Y)(q + wk_2 - \beta q)$$
$$= q + wk_2 - \beta q + Y(\beta q - \alpha q + wk_1 - wk_2) \tag{1}$$

$$U_{B_2} = 0 \tag{2}$$

$$\bar{B} = XU_{B_1} + (1 - X)U_{B_2} = X[q + wk_2 - \beta q + Y(\beta q - \alpha q + wk_1 - wk_2)] \tag{3}$$

Then the replicator dynamics equation for the consumer to select the 'browse' strategy is:

$$F(x) = \frac{dX}{d_t} = X(U_{B_3} - \bar{B}) = X(1 - X)[q + wk_2 - \beta q + Y(\beta q - \alpha q + wk_1 - wk_2)] \tag{4}$$

Similarly, the expected benefit of two strategies of the advertisers as well as the average benefit of the advertiser group is:

$$U_{A_1} = X\left(p + \frac{\sqrt{2}}{2}\lambda_1\alpha q\right) + (1 - X)\left[-\frac{1}{2}(\alpha q)^2\right] = -\frac{1}{2}(\alpha q)^2 + X\left(p + \frac{\sqrt{2}}{2}\lambda_1\alpha q\right) \tag{5}$$

$$U_{A_2} = X\left(p + \frac{\sqrt{2}}{2}\lambda_2\beta q\right) + (1 - X)\left[-\frac{1}{2}(\beta q)^2\right] = -\frac{1}{2}(\beta q)^2 + X\left(p + \frac{\sqrt{2}}{2}\lambda_2\beta q\right) \tag{6}$$

$$\bar{A} = Y\left[-\frac{1}{2}(\alpha q)^2 + X\left(p + \frac{\sqrt{2}}{2}\lambda_1\alpha q\right)\right] + (1 - Y)\left[-\frac{1}{2}(\beta q)^2 + X\left(p + \frac{\sqrt{2}}{2}\lambda_2\beta q\right)\right] \tag{7}$$

Then the advertiser chooses the replicator dynamics equation of 'social media advertising' as:

$$F(Y) = \frac{dY}{d_t} = Y(1 - Y)\left[X\left(\lambda_2 p - \lambda_1 p + \frac{\sqrt{2}}{2}\beta q - \frac{\sqrt{2}}{2}\alpha q\right) + \frac{1}{2}(\alpha q)^2 - \frac{1}{2}(\beta q)^2\right] \tag{8}$$

According to the replication dynamic system, the dynamic equilibrium solution is derived from the following equation:

$$\begin{cases} F(x) = 0 \\ F(Y) = 0 \end{cases} \tag{9}$$

When F(x) = 0, it is solved that X = 0, X = 1, Y = Y^*, among them, $Y^* = \frac{-(q+wk_2-\beta q)}{\beta q-\alpha q+wk_1-wk_2}$; when F(Y) = 0, it is solved that Y = 0, Y = 1, X = X^*, among them, $X^* = \frac{1}{2}\frac{(\beta q)^2-(\alpha q)^2}{\frac{\sqrt{2}}{2}\beta q\lambda_2-\frac{\sqrt{2}}{2}\alpha q\lambda_1}$. Therefore, the five equilibrium points in Eq. (1) are solved as O (0, 0); A (0, 1); B (1, 1); C (1, 0); D (X^*, Y^*).

According to the replicator dynamics equations of consumers and advertisers, this paper describes the replicator dynamics diagram about consumers and advertisers, as shown in Fig. 1 and Fig. 2.

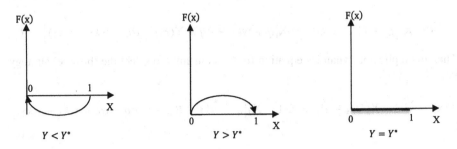

Fig. 1. Consumers replicator dynamics diagram

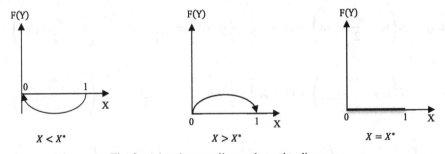

Fig. 2. Advertisers replicator dynamics diagram

Proposition 1. In the dynamic market, advertisers choose to invest social media advertising and consumers choose to browse, which are the evolutionary stability strategies. The evolution process is shown in Fig. 3.

As shown in Table 2, (1, 0) and (0, 1) belong to stable points among 5 equalization points, (0, 0), (1, 1) and (X^*, Y^*) belong to unstable points. In details, (1, 0) indicates that consumers browse social media advertisings; (0, 1) indicates that consumers do not browse the website advertisings. To achieve the biggest benefit, advertisers are more likely to invest social media advertisings as the final evolutionary stability strategy.

Figure 4 depicts the evolution process of consumers and advertisers. Assuming that D is the initial point, the area of quadrilateral AODB indicates the probability that the initial point finally stabilized at (1, 0). Similarity, the area of quadrilateral CODB indicates the probability that initial point will eventually stabilized at (0, 1).

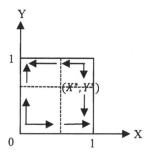

Fig. 3. Evolution stability legend

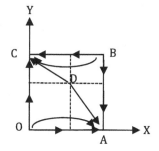

Fig. 4. Evolution of process by consumers and advertisers

Table 2. Stability analysis of equilibrium points

Equilibrium point	Det(J)	Tr(J)	Det(J)* Tr(J)	Stable or not
O (0, 0)	<0	<0	>0	Unstable
A (0, 1)	<0	>0	<0	stable
B (1, 1)	>0	>0	>0	Unstable
C (1, 0)	>0	<0	<0	stable
D (X^*, Y^*)	0	0	0	Unstable

Proposition 1 shows that when there is a difference between two channels of display advertisings in the attention cost contributed by consumers (i.e. $\alpha q < \beta q$), the higher cost of the consumers have paid, the more benefits that advertisers could gained. Therefore, advertisers are more likely to choose social media advertisings. Further more, if there is a difference between two channels of display advertisings which persuasive knowledge that could be activated to consumers (i.e. $k_1 w > k_2 w$), advertisers are more likely to choose social media advertisings with less activation. To sum up, obviously, compared with website advertising, social media advertising is more beneficial to advertisers. To deeply discuss the influence of consumers' factors on stable strategy, parameter analysis would be described in the following part.

5   Parameter Analyses

Proposition 2. Attention cost paid by consumers when they are browsing advertisings positively affects the probability that advertisers invest social media advertising ultimately. Specifically speaking, (1) if social media advertising costs consumers much more attention, advertisers are more likely to invest social media advertising; (2) if website advertising costs consumers' much more the attention, advertisers are more likely to invest website advertising.

Proposition 2 shows that when consumers pay more attention on social media advertising, market is more easy to invest social media advertising; When consumers pay more attention on website advertising, market is more difficult to stabilize to social media advertising strategy. Thus, compared with website advertising, consumers could get more information from social media advertising, so that social media advertising needs more attention from the individual. If consumers have paid such attention on browsing social media advertisings, they could be more likely to follow the ideas from the contents. To sum up, it supposes that advertisers tend to invest social media advertising.

Proposition 3. Persuasive knowledge which is activated by advertising is negatively affects the probability that advertisers invest social media advertising ultimately. Specifically speaking, (1) greater of the activation of consumer persuasive knowledge by website advertisings with the number of k_1, the greater probability that consumers and advertisers will choose social media advertisings; (2) smaller of the activation by social media advertising with the number of k_2, the greater probability that consumers and advertisers will choose social media advertising.

Proposition 3 shows that the persuasive knowledge that activated by advertisings is a crucial factor affecting the probability of the initial point eventually stabilizes to social media advertising. In details, when persuasive knowledge is activated more by website advertising, it is obvious that consumers may avoid website advertising, which makes it difficult for advertisers to invest website advertising; when persuasive knowledge is activated less by website advertising, it is suggested that advertisers should invest social media advertising.

Proposition 4. The browsing probability of channel is positively affects the probability that advertisers invest social media advertising ultimately. Specifically speaking, (1) compared with social media channel, if the browsing probability of website channel is higher, the market will be less likely stable to social media advertising; on the contrary, if the browsing probability of website channel is lower, the market will be more likely stable to social media advertising; (2) compared with website channel, if the browsing probability of social media channel is higher, the market will be less likely stable to social media advertising; on the contrary, the browsing probability of social media channel is lower, the market will be more likely stable to social media advertising.

Consequently, it could be noticed that the browsing probability of channel is one of the factors affecting the choice of advertising strategies. Considering a situation with the same pay-attention level, consumers will reject to browse such advertising that they find in with lower browsing probability of channel; while consumers will willing to browse such advertising which they find in with higher browsing probability of channel.

6   Data Simulate

In June 2017, a well-known Chinese insurance e-commerce platform (www.m.xyz.cn), tentatively decided to cancel search engine advertisings and invest social media advertising instead. It is found that the conversion rate of website advertisings is lower than that of social media advertisings when analysis from the purchase data from 2014 to 2017. This part simulates this transformation and analyzes the evolution results in detail using MATLAB. After discussion with managers, the following initial values of parameters which based on the model assumptions are proposed: $\alpha = 0.1$, $\beta = 0.3$, $k_1 = 0.3$, $k_2 = 0.1$, $q = 4$, $w = 5$, $\lambda_1 = 0.1$, $\lambda_2 = 0.3$. It also assumes that $X = 0.5$ is the browsing probability of consumers. Figures below have shown the impact of attention cost coefficient, persuasive knowledge activation coefficient as well as channel's influence on the result of evolutionary, when the initial probability of investing social media advertising are different. (When $Y = 0.2$, $Y = 0.4$, $Y = 0.6$, $Y = 0.8$, respectively).

(1) The effect of initial probability of publishing social media advertisings on evolutionary results. As shown in figures below, X and Y represent the initial behavior ratios of consumers and businesses in choosing website advertising and social media advertising, respectively. When initial behavior proportion of consumers keeps $X = 0.5$ all the time, the speed of evolution is consistent with the initial invest proportion of advertisers. The speed of evolution will increase when Y is becoming larger, and the speed is going to a peak when Y is close to the equilibrium point. Thus, although the value of Y is different, the final result of evolution is consistent, which also shows that the correlation between the advertisers' preference for investing social media advertising and the stable results is not significant.

(2) The effect of consumer attention cost coefficient on evolutionary results. Figure 5 depicts the impact of varies in attention cost paid by consumers when browsing social media advertisings on the evolution of the system, as the same time the attention cost coefficient of the consumers required to pay for website advertisings keeps constant. According to the system evolution paths in Fig. 5 (a) and Fig. 5 (b), it is found that when the attention cost coefficient of consumers browsing social media advertising is larger than that of browsing website, the system constrains to social media advertisings. However, with the increase of the attention cost coefficient of social media advertisings, the speed of constrains gradually decreases. Until beta = 0.7, the system began to constrain to website advertising, as shown in Fig. 5 (c). This change shows that the attention cost coefficient is not the higher, the better. When social media advertising needs to capture more and more consumers' attention, the market will eventually choose website advertising. This may be because too heavy advertising content may cause consumer displeasure, and too long length of space will also cause huge advertising costs for advertisers. Therefore, when publishing social media advertising, advertisers should pay attention to reasonably control the complexity of advertising content, so as to avoid making it difficult for consumers to understand and make the advertising effect counterproductive.

(3) The effect of persuasive knowledge activation coefficient on evolution result. Figure 6 depicts the impact of persuasive knowledge activated by website advertisings on the evolution result when the activation of persuasive knowledge by social media advertisings is constant. The more persuasive knowledge is activated, the quicker the process of constrain is. According to the evolution path of Fig. 6 (a), Fig. 6 (b) and Fig. 6 (c), it is found that when the website advertising activates more persuasive knowledge than social media advertising, the system eventually constrains to social media advertising. From this, we could notice that the degree of activation of persuasive knowledge in advertising has an important impact on the final convergence of the market. Therefore it is crucial for advertisers avoid such direct content to activate consumers' excessive persuasive knowledge. At the

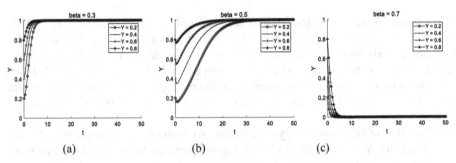

Fig. 5. The effect of consumer attention cost coefficient on evolutionary results. Figure (a) (b) and (c) represent the evolutionary results when $\beta = 0.3$, $\beta = 0.5$ and $\beta = 0.7$

same time, when k_1 takes different values, the constraining rate of the system does not change significantly. This conclusion also suggests that advertisers do not need to pay too much attention to the consumer persuasive knowledge stimulated by advertising.

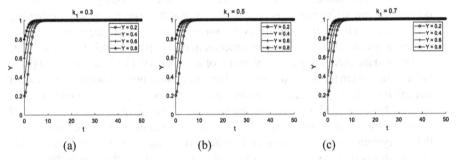

Fig. 6. The effect of persuasive knowledge activation coefficient on evolution result. Figure (a) (b) and (c) represent the evolutionary results when $k_1 = 0.3$, $k_1 = 0.5$ and $k_1 = 0.7$

(4) The influence of browsing probability on the evolution results. Figure 7 describes influence about the browsing probability of website channel on system evolution when the browsing probability of social media channel is constant. According to the evolution path of Fig. 7 (a), when the browsing probability of social media is greater than that of website, the system constrains to social media advertising. According to the evolution path of Fig. 7 (b) and Fig. 7 (c), it is obvious that when the browsing probability of social media is less than that of website, the system begins to converge to website and when $\lambda_1 = 0.5$ the convergence speed is obviously faster than $\lambda_1 = 0.3$. This conclusion also suggests that advertisers should attach importance to the influence of browsing probability on the choice of advertising channel. When the browsing probability is equal to or greater than that of social media, the system still converges to the website advertising even if attention coefficient of consumers is greater than that of website advertising, and the activation of persuasive knowledge is less than that of website advertising. Consequently, besides focusing on the editing of advertising content, advertisers

should also pay attention to the browsing probability of advertising channel and its impact on product brand.

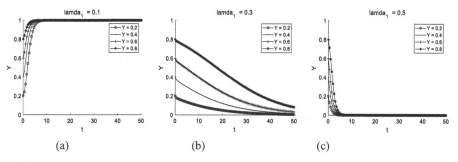

Fig. 7. The influence of browsing probability on the evolution results. Figure (a) (b) and (c) represent the evolutionary results when $\lambda_1 = 0.3$, $\lambda_1 = 0.5$ and $\lambda_1 = 0.7$

7   Results and Discussion

Based on consumer attention cost and persuasive knowledge, this paper constructs an evolutionary game model of consumers' advertising browsing behavior and advertisers online display advertising strategies. The model first considers that social media advertising could be gained more and more concerned by consumers according to the limited attention.

In details, website advertising with the characteristics of single-track or breakout in the early days (such as 'pop-up advertising'), interrupt the tasks that consumers are processing. While social media advertising enables consumers could have a conversation with advertisers directly, as well as discuss, cooperate and share advertising information with friends [21]. This 'participation' behavior captures a large amount of consumer's main attention, thus social media advertising will be more concerned by consumers. However, some scholars believe that social media advertising requires a lot of main attention, consumers may not willing to spend too much time to browse, as it described in Fig. 5 (c). Even so, social media advertising is still addicted consumers for its socially characteristic. This conclusion is consistent with Tutaj's suggestion that social media advertising are more attractive than traditional banner advertisings [22]. In the future, in the market where customers do not have a fixed brand preference, advertisers should try to elaborate advertising content that is more attractive to target users, as well as promote consumer pay much more attention [19].

In addition, the existing marketing conclusions indicate that the advertisers should integrate the strategic considerations of the advertising channel and the advertising channel, as well as the manager should fully consider the differences between channels before investing the advertising [23]. Thus, this paper suggests that advertisers could adopt different advertising strategies for different medias. For instance, advertisers could place more social media advertisings on such social media channels with accurately targeting and higher advertising effect. In the meanwhile, users are numerous, and advertisers could invest website advertising in the traditional website channel with a widely use of consumers and gained more requirement.

Secondly, the activation of persuasive knowledge has arouse the concern by the advertisers on the consumers psychology: the more obvious the meaning of the advertisings to be expressed by the advertisers, consumers are less likely to browse the advertisings. Once the advertiser realizes that the consumer would be less disrupt about the social media advertising, advertisers will actively invest the social media advertising. For example, users could notice the interactive content of the other online friends when they are browsing the social media advertising of Facebook. Such kind of interaction has constructed a situation combined with consumers, friends and brands, as well as enhances the consumer's social presence while reducing advertising dissatisfaction [24]. Therefore, advertisers should avoid direct advertising information in the design of advertising content, implicitly reveal the purpose of advertising, in order to dispel the tension of consumers and thus generate pleasure in advertising content [25].

The model also shows that advertisers will benefit more when the browsing probability of the channel is higher. The factors of the browsing probability of the channel involves many aspects, such as Ha et al. examined the mobile social media channel influence are concluded with individual perception and behavior [26]. In this study, social media channel is characterized with higher entertainment and lower ignorance, thus its browsing probability is higher evaluated. However, users alert to their privacy leaks with the differentiated push and accurate delivery of social media advertising. As the model shows, social media advertising effectively utilize the interactive features of social channels to promote interaction between consumers and channels, consumers and consumers while advertising, so as to enhance consumers' trust in advertising channels and reduce the negative emotions of consumers when browsing advertisings [19].

8 Summary

Prior studies have suggested that the psychological factors of consumers will affect the choice of advertising strategies. This paper describes the dynamic evolution process of consumers' browsing advertising behavior and advertisers choosing on online display advertisings through evolutionary game model, and explains the advertiser investment in dynamic market. The results show that: (1) the attention paid by consumers when browsing advertisings does have a significant impact on the advertisers' advertising strategy; (2) the activation of the consumer's persuasive knowledge also affects the choice of the advertisers' advertising strategy; (3) the influence of advertising channel also remarkably affects the decision-making of advertisers.

Compared with previous researches, the innovation of this paper is to describe the dynamic behavior of consumers' browsing advertising and advertisers online display advertising strategies through evolutionary game, while taking limited attention and persuasive knowledge of consumers into account. The final conclusion suggests that social media advertising will continue to be an evolutionary trend for advertisers to invest. This conclusion is consistent with the reality. The existing data shows that the channel of social media advertising has gradually extended to the search engine channel and short video channel in addition to news and video. Therefore, more advertisers could invest social media advertising to increase the profit based on the limited attention and persuasive knowledge activation of consumers. While it should be pointed out that this article does not consider the factors that affect the clicks behavior of advertisings, such as the size of the advertising title, the location of the advertising on the website, etc. These issues will be studied in the future research.

Acknowledgements. This research is supported by the National Natural Science Foundation of China under Grants (No. 71771122).

References

1. Soho news. https://www.sohu.com/a/246870763_99950158. Accessed 20 Dec 2019
2. Ghose, A., Todri, V.: Towards a digital attribution model: measuring the impact of display advertising on online consumer behavior. J. MIS Q. **40**, 889–910 (2016). https://doi.org/10.2139/ssrn.2638741
3. Manchanda, P., Dube, J., Goh, K., Chintagunta, P.: The effect of banner advertising on internet purchasing. J. Market. Res. **43**, 98–108 (2006). https://doi.org/10.1509/jmkr.43.1.98
4. Moe, W., Fader, P.: Dynamic conversion behavior at E-commerce sites. J. Manage. Sci. **50**, 326–335 (2004). https://doi.org/10.1287/mnsc.1040.0153
5. Xu, L., Duan, J., Whinston, A.: Path to purchase: a mutually exciting point process model for online advertising and conversion. J. Manage. Sci. **43**, 1392–1412 (2014). https://doi.org/10.2139/ssrn.2149920
6. Raney, A., Arpan, L., Pashupati, K.: At the movies, on the Web: an investigation of the effects of entertaining and interactive Web content on site and brand evaluations. J. J. Interact. Market. **56**, 38–53 (2010). https://doi.org/10.1002/dir.10064
7. Fransen, M., Verlegh, P., Kirmani, A.: A typology of consumer strategies for resisting advertising, and a review of mechanisms for countering them. J. Int. J. Advertising **34**, 6–16 (2015). https://doi.org/10.1080/02650487.2014.995284
8. Panic, K., Cauberghe, V., De, P.: Comparing TV Ads and advergames targeting children: the impact of persuasion knowledge on behavioral responses. J. Advertising **42**, 264–273 (2013). https://doi.org/10.1080/00913367.2013.774605
9. Zhu, Y., Wilbur, K.: Hybrid advertising auctions. J. Manage. Sci. **30**, 249–273 (2009). https://doi.org/10.2139/ssrn.1371758
10. Arnosti, N., Beck, M., Milgrom, P.: Adverse selection and auction design for internet display advertising. J. Am. Econ. Rev. **106**, 2851–2866 (2016). https://doi.org/10.1257/aer.20141198
11. Turner, J.: The planning of guaranteed targeted display advertising. J. Oper. Res. **60**, 18–33 (2012). https://doi.org/10.1287/opre.1110.0996
12. Zhang, K., Katona, Z.: Contextual advertising. J. Market. Sci. **31**, 980–994 (2012). https://doi.org/10.1287/mksc.1120.0740
13. Montgomery, A., Li, S., Srinivasan, K., Liechty, J.: Modeling online browsing and path analysis using clickstream data. J. Market. Sci. **23**, 579–595 (2004). https://doi.org/10.1287/mksc.1040.0073
14. Park, Y., Fader, P.: Modeling browsing behavior at multiple websites. J. Market. Sci. **23**, 280–303 (2004). https://doi.org/10.1287/mksc.1040.0050
15. Li, B., Yin, S., Xing, Z., Luo, X.: Coordinated development policy of energy storage industry in china based on evolutionary game theory. J. Ind. Eng. Manage. **24**, 171–179 (2019). https://doi.org/10.19495/cnki.10075429.2019.03.022

16. Liu, X., Sun, X., Wu, S.: Evolution of carbon emision game under dual governance system: analysis from the perspective of initial wilingnes diferentiation. J. Syst. Eng. **37**, 35–51 (2019)
17. Wu, J., Che, X., Sheng, Y.: Study on government-industry-university-institute collaborative innovation based on tripartite evolutionary game. J. Chin. J. Manage. Sci. **27**, 165–176 (2019). Doi:0.16381/j.cnki.isn1003-207x.2019.01.016
18. Ou.: Online advertising and pricing strategy based on evolutionary game. J. Manage. Rev. **35**, 181–187 (2015). https://doi.org/10.14120/j.cnki.cn11-5057/f.2015.06.018
19. Lee, D., Hosanagar, K., Nair, H.: Advertising content and consumer engagement on social media: evidence from Facebook. J. Manage. Sci. **64**, 5105–5131 (2018). https://doi.org/10.1287/mnsc.2017.2902
20. Schoonbeek, L., Kooreman, P.: The impact of advertising in a duopoly game. J. Int. Game Theory Rev. **09**, 565–581 (2007). https://doi.org/10.1142/S0219198907001606
21. Voorveld, H., Van, N., Muntinga, D.: Engagement with social media and social media advertising: the differentiating role of platform type. J. Advertising **47**, 38–54 (2018). https://doi.org/10.1080/00913367.2017.1405754
22. Lambrecht, A., Tucker, C.: When does retargeting work? Information specificity in online advertising. J. Market. Res. **50**, 561–576 (2013). https://doi.org/10.2139/ssrn.1795105
23. Wu, S., Hong, R., Jiang, L., Zhang, X.: Direct expression or indirect transmission? An empirical study on the impacts of metaphors and consumer's involvement level on advertising. J Bus. Rev. **29**, 133–142 (2017). https://doi.org/10.14120/j.cnki.cn11-5057/f.2017.09.012
24. Junglas, I., Goel, L., Abraham, C., Ives, B.: The social component of information systems—how sociability contribute to technology acceptence. J. Assoc. Inf. Syst. **14**, 585–616 (2013). https://doi.org/10.17705/1jais.00344
25. Yu, H., Chen, X.: Implicit or explicit: literature review and prospects of research on the effects of advertising metaphors. J. Foreign Economies Manage. **40**, 54–65 (2018). https://doi.org/10.16538/j.cnki.fem.2018.10.005
26. Ha, Y., Park, M., Lee, E.: A framework for mobile SNS advertising effectiveness: user perceptions and behaviour perspective. J. Behav. Inf. Technol. **33**, 1333–1346 (2014). https://doi.org/10.1080/0144929X.2014.928906

Platform Discount Deciding, Seller Pricing and Advertising Investment in the Shopping Festival Based on Two-Sided Market Theory

Hua Zhang, Li Li$^{(\boxtimes)}$, Xiang He, and Xingzhen Zhu

School of Economics and Management, Nanjing University of Science and Technology, Nanjing, China
Lily691111@126.com

Abstract. The online shopping platform, known as a two-sided market with buyers and sellers, always sponsor a Shopping Festival, such as Amazon "Black Friday", Taobao "Double 11". In this paper, a Stackelberg game model is constructed to study the platform discount deciding, seller pricing and advertising investment in the Shopping Festival. The research results show that (1) the optimal discount of the platform is related to the product buyer utility and unit cost. The higher the buyer utility of the product is, the larger the optimal discount coefficient of the platform is. The higher the unit cost of the product is, the larger the optimal discount coefficient for the platform is. (2) The platform profit increases in the number of buyers and the buyer utility of the product, and the profit increases in the platform transaction rate and the advertising cost coefficient firstly and then remains unchanged. (3) When the discount is more, seller's optimal pricing remains unchanged, and the advertising investment level reduce with the decrease of the platform discount coefficient. When discount is less, seller's optimal product pricing decreases in the decrease of the discount coefficient, but the advertising investment level remains unchanged.

Keywords: Two-sided market · Online shopping platform · Shopping Festival · Discount

1 Introduction

Online shopping platform could achieve the surge in transaction volume by sponsoring a Shopping Festival. For example, the transaction of Amazon's "Black Friday" Shopping Festival amounted to 6.2 billion dollars on November 23, 2018. The transaction of Taobao "Double 11" Shopping Festival amounted to 213.5 billion yuan on November 11, 2018. The online shopping platform with two-sided market has cross-network externalities. Sellers prefer to participate in the Shopping Festival with many buyers, and buyers prefer to participate in the platform with more sellers to have more product selection, and prefer discounts to gifts [1]. Therefore, the platform sponsors the Shopping Festival by using discount as a promotional tool. We take the Taobao "Double 11" as an example of the Shopping Festival. Firstly, Taobao announces discount in "Double 11", the discount is

© Springer Nature Switzerland AG 2020
K. R. Lang et al. (Eds.): WeB 2019, LNBIP 403, pp. 103–114, 2020.
https://doi.org/10.1007/978-3-030-67781-7_10

borne by sellers. Then, sellers sell products according to the discount if they participate in the Shopping Festival. Less discount is difficult to attract buyers, and insufficient buyers participating in the Shopping Festival will not attract more sellers to participate in the Shopping Festival. More discount attracts more buyers and affects sellers' profits and reduce the willingness of sellers to participate in the Shopping Festival. Therefore, for the online shopping platform, how to determine reasonable discount to attract sellers and buyers to participate in the Shopping Festival is critical.

The discount decision belongs to the research field of two-sided platform pricing. Rochet and Tirole [2] and Armstrong and Wright [3] studied the pricing of two-sided platform earlier and expressed that the two-sided platform provides services and sets rules of transaction, then two-sided users access the platform, match and trade on the platform. The pricing strategy of the platform will affect the entry of two-sided users and the overall demand scale of the platform [3]. Some scholars have explored the main factors affecting the pricing strategy of two-sided platforms. Rochet and Tirole [4] analyzed the impact of two-sided user demand price elasticity on the pricing of platform transaction fees in the study of transaction pricing of two-sided platforms. The result shows that the optimal transaction cost of the platform is proportional to price elasticity. Belleflamme and Peitz [5] studied seller investment incentives under two competing two-sided platforms, the result shows that for-profit intermediation may lead to overinvestment when free access would lead to underinvestment. Reisinger [6] studied platform competition for advertisers and users in media markets. In addition, some scholars considered the impact of seller independent decision-making behavior on platform pricing strategies. Hagiu [7] studied the pricing of membership fees in two-sided platforms considered the product price decision of seller, and established a three-stage game model. Firstly, the platform makes the pricing of membership fees. Secondly, the seller and buyer decided whether to join the platform. Lastly, seller makes pricing of the products, and the buyer decides whether to buy. Cao and He [8] explored the pricing competition of B2C platform and third-party seller. In the above papers, the single decision of seller on product pricing investment is considered.

This paper established a Stackelberg model to study the relationship between online shopping platform and seller in the Shopping Festival. (1) We studied the market mature platform in the Shopping Festival, the source of the online shopping platform profit does not include membership fees but includes seller profit-sharing and advertising fees, and it is free for buyers to enter to platform. (2) Seller considers the decision of pricing and advertising investment. (3) The discount is different from the direct pricing of fees. On the one hand, the platform discount is the promise of discount to buyers. On the other hand, the platform discount borne by seller is the prerequisites of seller entering the Shopping Festival, such as "full 400 minus 50" during the Taobao "Double 11". Through analysis of equilibrium, we studied the online shopping platform discount, seller pricing and advertising investment in the Shopping Festival, to provide theoretical support for the improvement of transaction efficiency in the e-commerce market.

2 Model

Considering a completely monopolistic two-sided market platform, sellers and consumers are two-sided users of the platform. The platform charges seller transaction fees and advertising service fees [8], which is free to consumers. The model has two stages. In the first stage, the platform as the leader determines the discount for the Shopping Festival. In the second stage, the seller as a follower determines whether to participate in the Shopping Festival by the analysis of the discount published. Participating in the Shopping Festival, the seller will determine the price of the product and the level of advertising investment. If not, the seller also needs to determine the price of the product, because the seller has the demand in the natural market, the product can also be retrieved and viewed by consumers on the online shopping platform, in this time, seller often needs to determine a lower price to attract consumers to buy.

Consumers are risk-neutral, and they decide whether to purchase according to the perceived utility. When the perceived utility is greater than zero, the consumer decides to purchase, otherwise, does not purchase. Consumer perceived utility is influenced by many factors such as product quality factors and product services. Let U denote the consumer's perceived utility for the product, and φ is the reserve price that expresses the maximum consumer's perceived utility when the product price is zero. Before the Shopping Festival, the product price is p_0, at which point the consumer perceived utility is $U_0 = \varphi - p_0$. Consumers who get the advertisement of the product have the corresponding utility perception for the product, at the same time, consumers get the information about prices of product and platforms discount. If the perception of consumer utility $U \geq 0$, consumers choose to buy the product, otherwise not to buy.

We assume that p_1 is the seller's product price if the seller participates in the Shopping Festival, and the platform stipulates that the seller who participates in the Shopping Festival can only reduce the price, but cannot raise the price, that is, $p_0 \geq p_1$. We assume that p_2 is the seller's product price if the seller does not participate in the Shopping Festival, because the seller adjusts the price of the product to p_2 in response to changes in consumer demand in the platform. In the Shopping Festival, the quality, functions, and services of the product are not affected by the Shopping Festival, but the consumption is delayed due to the increase in the sales volume of the Shopping Festival of the platform. With the logistics service level is reduced, the satisfaction of the purchased product is reduced, and the consumer perceived utility is reduced to $\gamma\varphi$, where γ is the consumer perceived utility coefficient in the Shopping Festival, $0 < \gamma < 1$. In this time, the consumer purchases product that participates in the Shopping Festival, and the consumer perceived utility:

$$U_1 = \gamma\varphi - kp_1 \tag{1}$$

If the consumer buys a product that does not participate in the Shopping Festival, then the consumer perceived utility:

$$U_2 = \gamma\varphi - p_2 \tag{2}$$

If the seller participates in the Shopping Festival, the seller has the right to be provided the advertising service by the platform. Assume that advertisements are only informative and not persuasive, that is, advertisements only convey product information and do not change consumer preferences, that is, do not change consumers' own retention prices [9]. Let a denote the advertisement level corresponding to the seller advertising investment, $a \in (0, 1)$, and it expresses that the consumer in the platform receives the advertisement with the probability a. The fees of the advertisement level a is $fa^2/2$, and f is the advertisement fees coefficient [10]. If the seller does not participate in the Shopping Festival, the seller cannot get the advertising service provided in the Shopping Festival, so the seller has the natural market without advertisement, and consumers get the product information according to the probability of the original natural market a_0. In other words, consumers in the natural market have a perceived utility when they get the product information, but other consumers outside the natural market do not receive the product information, so they have no perceived utility.

Observing consumers' behavior, the seller determines the advertising level for the product. We assume that consumers' perception of the utility of the product is heterogeneous. Note that, U is uniformly distributed over $[0, \varphi]$. Seller advertising level is a, therefore the expected number of consumers receiving ads is an_c. In the number of an_c consumers, consumers whose perception of product utility is more than zero will buy it. The consumer demand curve is shown in Fig. 1:

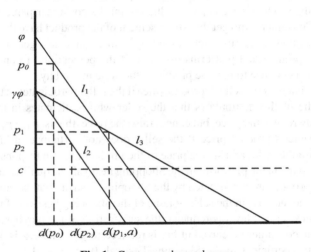

Fig. 1. Consumer demand curve

In Fig. 1, l_1 represents the consumer demand curve before the Shopping Festival. Consumers have product information based on the probability of the natural market. When the price is p_0, the demand for the product is $d(p_0)$. l_2 represents the consumer demand curve when the seller does not participate in the Shopping Festival. The logistics lag and service levels in the Shopping Festival make the perceived consumer utility become $\gamma\varphi(c < \gamma\varphi < \varphi)$, so the consumer demand curve from l_1 to l_2. l_3 represents the consumer demand curve of sellers participating in the Shopping Festival. During the Shopping Festival, the seller advertising levels for a, the number of consumers receive advertising is an_c. We assume that advertising is informative but not persuaded, so receiving the advertisement, consumers' utility remains unchanged, so the demand curve from l_2 to l_3. According to consumer demand curve l_3 is shown in Fig. 1, if the seller participates in the Shopping Festival, the number of consumers receiving ads is an_c when seller advertising level is a, and the slope of the demand curve for $-\varphi/an_c$, the demand curve expression is $kp_1 = -\frac{\varphi}{an_c}d(kp_1, a) + \gamma\varphi$, by formula conversion available:

$$d(kp_1, a) = \frac{an_c(\gamma\varphi - kp_1)}{\varphi} \tag{3}$$

Then, we have:

$$d(p_2) = \frac{a_0 n_c(\gamma\varphi - p_2)}{\gamma\varphi} \tag{4}$$

Let Π_{s1} denote seller's profit when seller participates in the Shopping Festival. That is,

$$\Pi_{s1} = (1 - \theta)(kp_1 - c)d(kp_1, a) - \frac{1}{2}fa^2 \tag{5}$$

Where $d(kp_1, a)$ represents the demand when the platform discount is k, the advertising level a, and the product price is p_1 in the Shopping Festival. The first part of the formula (5) is seller sales profit, the second part is the advertising fees paid to the platform for seller. Let Π_{s2} denote seller's profit when seller does not participate in the Shopping Festival. That is,

$$\Pi_{s2} = (1 - \theta)(p_2 - c)d(p_2) \tag{6}$$

Where $d(p_2)$ denote the buyer demand of seller in the natural market.

By predicting seller behavior and buyer behavior, the platform analyzes the platform's maximum profit. Sponsoring a Shopping Festival, the platform benefits are Π_p. That is,

$$\Pi_p = \theta(kp_1 - c)d(kp_1, a) + \frac{1}{2}fa^2 \tag{7}$$

3 Equilibrium Analysis

3.1 Seller Optimal Decisions

If seller participate in the shopping festival, we have Proposition 1. (Note: proof of proposition in this paper is omitted)

Proposition 1: When seller participates in the Shopping Festival, the optimal price is

$$p_1^* = \begin{cases} \frac{c+\gamma\varphi}{2k}, & \frac{c+\gamma\varphi}{c+\varphi} \le k < 1 \\ \frac{c+\varphi}{2}, & \frac{2c}{c+\varphi} \le k < \frac{c+\gamma\varphi}{c+\varphi} \end{cases} \tag{8}$$

Seller optimal advertising level is

$$a^* = \begin{cases} \frac{n_c(1-\theta)(\gamma\varphi-c)^2}{4f\varphi}, & \frac{c+\gamma\varphi}{c+\varphi} \le k < 1 \\ \frac{n_c(1-\theta)(kp_0-c)(\gamma\varphi-kp_0)}{f\varphi}, & \frac{2c}{c+\varphi} \le k < \frac{c+\gamma\varphi}{c+\varphi} \end{cases} \tag{9}$$

Proposition 1 shows that optimal pricing increases in k. After k decrease to $(c + \gamma\varphi)/(c + \varphi)$, seller's optimal pricing remains the same value. When k is smaller than $2c/(c+\varphi)$, seller does not participate in the Shopping Festival. Thus, it can be seen that seller optimal pricing can only be adjusted when the discount coefficient of the platform is high. When the discount coefficient is low, the optimal pricing of seller participating in the Shopping Festival is the original price. The optimal pricing does not change with the size of the advertising level, that is, seller pricing decisions can be independent of advertising investment decisions. the optimal advertising level is a constant value of $\frac{n_c(1-\theta)(\gamma\varphi-c)^2}{4f\varphi}$ when k decreases from 1 to $\frac{c+\gamma\varphi}{c+\varphi}$. When k continues to decline to $\frac{2c}{c+\varphi}$, the optimal advertising level decreases.

Proposition 2: Participating in the Shopping Festival, seller profit is

$$\Pi_{s1} = \begin{cases} \dfrac{n_c^2(1-\theta)^2\left(\frac{\gamma\varphi-c}{2}\right)^4}{2f\varphi^2}, & \dfrac{c+\gamma\varphi}{c+\varphi} \leq k < 1 \\[4mm] \dfrac{n_c^2(1-\theta)^2\left(k\frac{c+\varphi}{2}-c\right)^2\left(\gamma\varphi-k\frac{c+\varphi}{2}\right)^2}{2f\varphi^2}, & \dfrac{2c}{c+\varphi} \leq k < \dfrac{c+\gamma\varphi}{c+\varphi} \end{cases} \tag{10}$$

When the discount coefficient is large, seller profit does not change with the discount coefficient. When the discount coefficient is small, the lower discount coefficient, seller profit decreases. Participating in the Shopping Festival, seller's profit increases in the number of buyers, the retention price of the product and the buyers' perceived utility coefficient. Seller's profit decreases with the advertising cost coefficient, transaction rate and unit cost of the product.

If seller does not participate in the shopping festival, the goods still have a market, which is the customer who has a certain amount of attention and knowledge of the goods, so seller will still be able to obtain higher returns by reducing the price, then, we have Proposition 3.

Proposition 3: When seller does not participate in the Shopping Festival, seller optimal pricing is

$$p_2^* = \frac{c + \gamma\varphi}{2} \tag{11}$$

we can find that the buyer perception of the utility coefficient is the main factors that influence the p_2, so lowering prices compensates to reduce the utility of buyer product in the Shopping Festival.

Proposition 4: When seller does not participate in the Shopping Festival, the optimal profit is

$$\Pi_{s2} = \frac{a_0 n_c(1-\theta)(\gamma\varphi - c)^2}{4\gamma\varphi} \tag{12}$$

When seller does not participate in the Shopping Festival, seller's profit increases in the probability that natural market buyers get information about a product, the number of buyers and the perceived utility of the product. Seller profit is reduced with the transaction rate of the platform and the cost of commodity units. The transaction rate of the platform is the factor that affects seller on entering the platform and the participation of the platform activities, and the higher the transaction rate of the platform, the higher the transaction fee of seller, and the lower the income of seller. The control of product unit cost affects seller profit directly.

Seller decides whether to participate in the Shopping Festival or not by comparing the profit of participating in the Shopping Festival with that of not participating in the Shopping Festival. When $\frac{2c}{c+\varphi} \leq k < \frac{c+\gamma\varphi}{c+\varphi}$, we have

$$\Pi_{s1} - \Pi_{s2} = \frac{n_c^2(1-\theta)^2\left(k\frac{c+\varphi}{2}-c\right)^2\left(\gamma\varphi - k\frac{c+\varphi}{2}\right)^2}{2f\varphi^2} - \frac{a_0 n_c(1-\theta)(\gamma\varphi - c)^2}{4\gamma\varphi}.$$

The above equation is simplified to solve the unitary quadratic equation about k. The equation is as follows: $-k^2 + \left(\frac{c+\gamma\varphi}{p_0}\right)k - \frac{c\gamma\varphi}{p_0^2} - \frac{(\gamma\varphi-c)}{p_0^2}\sqrt{\frac{a_0 f\varphi}{2\gamma n_c(1-\theta)}} = 0$. Firstly, we determine whether there is a real solution to the equation. If there is no real solution, it means that the profit of seller participating in the Shopping Festival is always smaller than that of seller not participating in the Shopping Festival. Only when there is a real solution can seller participate in the Shopping Festival. Then the discriminant is $\Delta = \left(\frac{c+\gamma\varphi}{p_0}\right)^2 - 4\left(\frac{c\gamma\varphi}{p_0^2} + \frac{(\gamma\varphi-c)}{p_0^2}\sqrt{\frac{a_0 f\varphi}{2\gamma n_c(1-\theta)}}\right)$. If Δ is less than zero, seller chooses not to participate in the Shopping Festival. If Δ is more than zero, solve the equation about k, the solution is $\frac{\gamma\varphi+c}{\varphi+c} - \frac{1}{2}\sqrt{\Delta} \leq k \leq \frac{\gamma\varphi+c}{\varphi+c} + \frac{1}{2}\sqrt{\Delta}$. If Δ is equal to zero, it means that there is a unique solution, which makes the profit of participating in the Shopping Festival the same as that of not participating in the Shopping Festival. That is $k = \frac{\gamma\varphi+c}{\varphi+c} - \frac{1}{2}\sqrt{\Delta}$. If $\frac{c+\gamma\varphi}{c+\varphi} \leq k < 1$, seller achieves the optimal benefits. If $\frac{2c}{c+\varphi} \leq k < \frac{c+\gamma\varphi}{c+\varphi}$, seller maybe takes part in the Shopping Festival. If $\frac{c+\gamma\varphi}{c+\varphi} \leq k < 1$, seller takes part in the Shopping Festival inevitable choice when Δ is greater than or equal to zero.

According to the discount, seller decides on whether to participate in the Shopping Festival. Through the above calculation, there is a threshold value, when the threshold value is less than zero, for any discount on the platform, the maximum benefit of seller who do not participate in the Shopping Festival is greater than the maximum profit of participating in the Shopping Festival, seller chooses not to participate in the Shopping Festival. When the threshold value is greater than or equal to zero, existing seller to participate in the Shopping Festival maximum profit greater than not to participate. When the platform discount coefficient is less than $\frac{\gamma\varphi+c}{\varphi+c} - \frac{1}{2}\sqrt{\Delta}$, seller chooses not to participate in the Shopping Festival. Then if $\frac{2c}{c+\varphi} < k < \frac{\gamma\varphi+c}{\varphi+c} - \frac{1}{2}\sqrt{\Delta}$, the profit of seller participating in the Shopping Festival is higher than that of non-participants when $0 < k < \frac{2c}{c+\varphi}$. If seller participates in the Shopping Festival, the actual selling price is lower than the product unit cost, so seller will lose money. When the platform discount is more than $\frac{\gamma\varphi+c}{\varphi+c} - \frac{1}{2}\sqrt{\Delta}$, seller chooses to participate in the Shopping Festival. If $\frac{c+\gamma\varphi}{c+\varphi} \leq k < 1$, seller achieves maximum benefits.

3.2 Platform Optimal Decisions

Proposition 5: Platform optimal discount coefficient is k^*, $\frac{c+\gamma\varphi}{c+\varphi} \leq k^* < 1$. The platform maximal profit is $\Pi_p^* = \frac{(1-\theta^2)n_c^2\left(\frac{\gamma\varphi-c}{2}\right)^4}{2f\varphi^2}$.

Proposition 5 shows that the optimal discount coefficient is not the lower the better, on the contrary, the optimal discount coefficient is to maintain a high level, that is, the platform has a small degree of discount. This is contrary to the practical experience because the lower discount coefficient of the platform enables the platform to get more transaction shares, but the reduced advertising investment from seller makes the platform's profit lower. If the platform discount coefficient is high, seller must find the optimal product demand curve to make optimal price, so the actual transaction price is lower than the discount. Although the price is lower, seller actual clinch a deal the price is higher, trading volume is reduced, but seller's advertising investment increase, making platform finally gains.

4   Analysis

Proposition 6: Platform optimal discount coefficient k is related to buyer perceived utility coefficient γ, unit costs c and buyers' reserve price φ. The lower boundary of the optimal discount coefficient k increases in buyer perceived utility coefficient γ and unit costs of product c. The lower boundary of the optimal discount coefficient k decreases with buyers' reserve price φ.

During Shopping Festival, buyer perceived utility coefficient γ expresses the degree to which factors such as logistics lag and service level decline in the Shopping Festival reduce buyers' perceived utility. The higher the value of φ is, the smaller the reduction of buyers' perceived utility is, and the higher the lower bound of the value of the platform's optimal discount is. When the platform reduces the negative impact of Shopping Festival on buyers, the Shopping Festival can attract buyers' attention with a low discount. Besides the cost of the product and buyers' perceptions utility, the platform should publish different discount for different sellers. The platform should set a high discount coefficient for the high-cost product because if the discount is too high, sellers' profit is lower, many sellers do not participate in the Shopping Festival. For the product with higher reserved prices by buyers, due to sellers can adjust the price in a larger range, the optimal discount is lower and the value range of the optimal price is larger.

Proposition 7: The larger the three values of buyer perceived utility coefficient, reservation price and the number of buyers, seller tends to participate in the Shopping Festival. The higher the three values of product cost, transaction rate, and advertising cost coefficient, seller tends not to participate in the Shopping Festival.

Seller's decision depends on the marginal profit of the product. The larger value of marginal profit of product is, more inclined seller is to participate in the Shopping Festival. The lower the cost and the higher buyer perceived utility, more inclined seller are to participate in the Shopping Festival. The smaller the negative impact, seller tends to take part in the Shopping Festival. For example, the platform ensures delivery speed, reduces the shortage risk during the Shopping Festival and guarantee the quality of product and services, then seller is more inclined to participate in the Shopping Festival. Also, the greater buyer utility values, more inclined seller is to participate in the Shopping Festival. It is generally believed that sellers whose product has low buyer utility tend to participate in the Shopping Festival, such as out-of-season clothing product. However, the conclusion shows that the bigger the buyer utility product more inclined to take part in the Shopping Festival, it is because that the price of the product with bigger buyer utility is higher than the product with small buyer utility. The lower price of the product with small buyer utility leads to lower profit. Therefore, the higher the utility of the product is, the more likely they are to participate in the Shopping Festival. The larger number of buyers in the platform also promotes seller to participate in the Shopping Festival. Meanwhile, during the Shopping Festival, if the advertising cost coefficient and the transaction rate are reduced, seller needs to pay a lower the advertising fees and transaction cost, and seller will get a higher profit, so they decide to participate in the Shopping Festival. The increase of a_0 will urge seller not to participate in the Shopping Festival, because when seller does not participate in the Shopping Festival, there is still a part of market, and the larger the market is, the greater the profits will be, so seller is more inclined not to participate in the Shopping Festival.

Proposition 8: When the platform discount is large, the maximum profit of seller decreases with the decrease in the value of k. When the value of k decreases to Δ, seller does not participate in the Shopping Festival. When the platform discount is small, the maximum profit of seller is independent on the value of k.

When the discount is small, seller participates in the Shopping Festival at the optimal actual transaction price, and the profit is Π_{s1}. When the discount is large, seller profit decreases with the decrease in the value of k. When the value of k decreases to Δ, the profit of seller participating in the Shopping Festival equal to the profit of seller not participating in the Shopping Festival. When the value of k continues to decrease, the profit of seller who does not participate in the Shopping Festival is more than that of seller who participate in the Shopping Festival, so seller decides not to participate in the Shopping Festival.

Proposition 9: When k is smaller than $\frac{\gamma\varphi+c}{\varphi+c} - \frac{1}{2}\sqrt{\Delta}$, the platform does not sponsor a Shopping Festival. When $\frac{\gamma\varphi+c}{\varphi+c} - \frac{1}{2}\sqrt{\Delta} \le k < \frac{c+\gamma\varphi}{c+\varphi}$, the platform sponsors the Shopping Festival, and platform profit increases in valve of k. When $(c + \gamma\varphi)/(c + \varphi) \le k < 1$, platform profit reaches the maximum, and the maximum value of the profit remains unchanged with the increase of k. The platform optimal discount range is $(c + \gamma\varphi)/(c + \varphi) \le k < 1$.

When the discount coefficient of the platform is kept at a high level, the platform gains the most, that is, during the Shopping Festival, the greater the discount, the greater the benefits. The reason is that the small discount coefficient leads to the lower price, according to the proof of Proposition 2, the advertising investment of seller also decreases accordingly.

Proposition 10: Platform profit increases in the number of buyers, buyer reserve price and buyer perceived utility coefficient. Platform profit decreases with transaction rate, advertising cost coefficient and unit cost of the product.

Accordingly, we can find that enough number of buyers is the base of the platform to sponsor a Shopping Festival. The equilibrium result shows that the reduction of transaction rate and advertising fees coefficient is conducive to the increase of platform profit, which is obviously not intuitive. From the practical experience, the reduction of transaction rate and advertising fees will reduce the transaction income and advertising fees of the platform, and it is obviously not conducive to the growth of platform profit. Results can be obtained from the model, however, when the transaction rates and advertising costs coefficient are reduced, seller's pricing is not affected, but seller optimal advertising level increases. While the lower transaction rates will reduce the platform transaction fees, but the platform's incremental advertising fees make up for the drop-in transaction fees, the platform total profit is still increasing. Although the platform profit of transaction fees and advertising fees coefficient is increased, according to the definition of advertising level, the value of seller advertising level is smaller than 1. Due to the limitation of the platform technology, seller's advertising level cannot continue to improve after reaching a certain level. At this time, if the platform continues to reduce the transaction rate and the advertising fees coefficient, seller is not likely to increase the continued advertising investment. So, the platform tends to maximize the advertising investment, rather than simply lowering the price. In addition, the platform enhances the level of sponsoring the Shopping Festival, reduces the lag of logistics caused by the Shopping Festival, and the risk of out-of-stocks, which adds buyers the buyer utility of the product, meanwhile the improvement of the Shopping Festival will benefit the platform, sellers and buyers. At last, improving buyer utility can also increase the profit of the platform, and the platform should strengthen the platform brand building.

5   Managerial Implications and Discussion

In the context of the Shopping Festival of the online shopping platform, we established a game model of a two-sided platform and a seller, analyzes the platform optimal discount, seller's product pricing and the advertising investment decision. Results can reveal the rationality of the existing phenomena in the online shopping platform management practice, and provide support for the decision of the platform and seller.

Firstly, the discount is not as big as possible. The larger discount will reduce the profit of sellers, thus reducing the advertising investment of sellers. On the contrary, the platform provides a smaller discount, which makes sellers have a certain price reduction space. Sellers will make the decision of pricing and advertising investment in the Shopping Festival according to the discounts released by the platform and the demand curve of the products, which can achieve "win-win" for the platform and sellers. The optimal discount is related to attributes of products and should be classified and promoted according to the different attributes of the products in the platform. The traditional platform unified discount has the following problems. First, the uniform discount will make sellers with low marginal profit not participate in the Shopping Festival. Second, if the platform's unified discount is less, buyers' shopping desire will be reduced. Therefore, the discount should be classified according to the unit cost of the product, the utility of the buyer and the degree of change in the buyer utility brought by the Shopping Festival to different products. Different discounts for different products will increase the number of sellers and buyers and improve platform profit. In addition, improving the logistics service level in the Shopping Festival and ensuring the quality of the product not only improve the welfare of buyers but also help to increase the profit of sellers and platforms. The platform profit increases in the number of buyers and the buyer utility of the product. The platform profit decreases with the increase of the platform transaction rate and the advertising cost factor. The platform will increase the national advertising level in the early stage of the Shopping Festival, and increasing the number of buyers participating in shopping will play an important role in improving platform profit. Reducing appropriately the advertising cost or subsidizing the advertising cost of the seller will increase the profit of the seller and the sales volume of the platform, thus increasing the profit of the platform.

Secondly, deciding on whether to participate in the Shopping Festival, seller not only must consider the discount of the platform but also needs to analyze multiple parameters such as the unit cost of the product, the utility of the buyer, the advertising cost and the transaction rate. Only if these parameters met a certain condition, seller has the motivation to participate in the Shopping Festival. The greater the buyer perceived utility coefficient, the retention price and the number of buyers, the more both profits of sellers participating in and not participating in the Shopping Festival are. Because the profit of seller participating in the Shopping Festival is more than that of not participating in, sellers tend to Participate in the Shopping Festival. The greater the unit cost and transaction rate of the product, the lower both profits of sellers participating in and not participating in the Shopping Festival are. In addition, with the increase of the advertising cost and the probability that the natural market buyers getting information, sellers prefer not to participate in the Shopping Festival. Improving the buyer utility and reserve level in the Shopping Festival, such as strengthening the investment in brand advertising, improving the quality of product and services, sellers can improve the profit. For sellers who do not participate in the Shopping Festival, in order to cope with changes in demand, reducing price can increase market demand, and having a stable natural market is a guarantee for sales of sellers.

Finally, pricing and advertising investment decisions are related to the discounts of the platform. When the discount is low, sellers have space for price adjustment. At this time, sellers determine the optimal pricing to obtain the maximum profit, and sellers actually adjust the product price as the discount of the platform effects the product price. The price remains the optimal price during the Shopping Festival, and the corresponding optimal advertising level does not change with the discount of the platform. When the discount is high, sellers choose to adjust the price, and sellers' profits is reduced. At this time, the price will remain unchanged. As the discount coefficient of the platform continues to decrease, the actual transaction price decreases, and the corresponding optimal advertising level also follows. Reduced.

6 Conclusion

In this research, the main conclusions are as follows: (1) The optimal discount of the platform is related to the utility of the product buyer and the unit cost. The higher the buyer utility of the product is, the larger the optimal discount coefficient is. And the higher the unit cost of the product is, the greater the discount coefficient is. (2) In the Shopping Festival, the platform profit increases in the number of buyers and the buyer utility of the product. With the increase of the platform transaction rate and the advertising cost coefficient, the platform profit will increase firstly and then remain unchanged. (3) If the platform discount coefficient is low, seller's optimal pricing remains unchanged with that price before the Shopping Festival, the advertising investment level as the platform discount coefficient decreases. If the platform discount coefficient is high, seller's optimal pricing increases in the platform discount coefficient, but the advertising investment level remains unchanged.

Acknowledgements. This research was supported by the National Natural Science Foundation of China under Grants 71771122.

References

1. Weisstein, F.L., Monroe, K.B., Kukarkinney, M.: Effects of price framing on consumers' perceptions of online dynamic pricing practices. J. Acad. Mark. Sci. **41**(5), 501–514 (2013). https://doi.org/10.1007/s11747-013-0330-0
2. Rochet, J., Tirole, J.: Platform competition in two-sided markets. J. Eur. Econ. Assoc. **1**(4), 990–1029 (2003). https://doi.org/10.1162/154247603322493212
3. Armstrong, M., Wright, J.: Two-sided markets, competitive bottlenecks and exclusive contracts. Econ. Theory **32**(2), 353–380 (2007). https://doi.org/10.1007/s00199-006-0114-6
4. Rochet, J., Tirole, J.: Two-sided markets: a progress report. RAND J. Econ. **37**(3), 645–667 (2006). https://doi.org/10.1111/j.1756-2171.2006.tb00036.x
5. Belleflamme, P., Peitz, M.: Platform competition and seller investment incentives. Eur. Econ. Rev. **54**(8), 1059–1076 (2010). https://doi.org/10.1016/j.euroecorev.2010.03.001
6. Reisinger, M.: Platform competition for advertisers and users in media markets. Int. J. Ind. Organ. **30**(2), 243–252 (2012). https://doi.org/10.2139/ssrn.1855287
7. Hagiu, A.: Two-sided platforms: Pricing and social efficiency. J. Econ. Manag. Strategy **18**(4), 1011–1043 (2009). https://doi.org/10.2139/ssrn.621461
8. Cao, K., He, P.: The competition between B2C platform and third-party seller considering sales effort. Kybernetes **45**(7), 1084–1108 (2016). https://doi.org/10.1108/K-01-2016-0009
9. Bagwell, K.: The economic analysis of advertising. Handb. Ind. Organ. **3**(06), 1701–1844 (2007). https://doi.org/10.7916/D8QZ2P6X
10. Butters, G.R.: Equilibrium distribution of sales and advertising prices. Rev. Econ. Stud. **44**(3), 465–491 (1977). https://doi.org/10.2307/2296902

Who Picks Cherries? Understanding Consumers' Cherry Picking Behavior in Online Music Streaming Services

Changkeun Kim, Byungjoon Yoo, and Jaehwan Lee[(⊠)]

Seoul National University, Seoul 08826, South Korea
jlee@idb.snu.ac.kr

Abstract. Subscription business model is common in many online services. It is crucial to service providers to acquire and retain more customers. Because customers can switch one service to another easily, price promotion may cause economic loss of the providers. The aim of this paper is to investigate the characteristics of cherry pickers in subscription-based online services. We find that cherry pickers use online music streaming services more actively and apply their investment to the service more readily than non-cherry pickers.

Keywords: Cherry picking · Price promotion · Deal proneness · Subscription online service

1 Introduction

In recent years, there has been an increase in the number of online services whose entire business relies on a subscription business model. These include Netflix, Spotify, and SaaS (Software as a service) companies. Subscription pricing is especially beneficial to streaming media providers. Cumulative costs for subscription pricing will soon be higher than the purchasing price, and this can support providers' maintenance expenses including network traffic and distribution fees.

In contrast to the offline environment, customers can switch their patronage among online services easily. If a customer wishes to switch one service to another, he or she needs only to stop purchasing the unwanted services and sign up for the new. To attract customers, many online services use a price promotion marketing strategy. They offer discounts for a short introductory period upon purchase. The problem is that, due to the ease of online service switching behavior, some customers take the promotion and leave after the promotion period has elapsed. Such customers who seek to exploit these deals are called "cherry pickers". An excess of cherry pickers could have a negative effect on the profit of the service providers.

Our paper aims to explain the characteristics of cherry pickers in terms of usage and investment in subscription based online services. Based on the existing literature on cherry pickers and price promotion, we build the following two hypotheses: (1) Cherry pickers use online subscription services more actively than non-cherry pickers do, and

© Springer Nature Switzerland AG 2020
K. R. Lang et al. (Eds.): WeB 2019, LNBIP 403, pp. 115–122, 2020.
https://doi.org/10.1007/978-3-030-67781-7_11

(2) Cherry pickers devote less investment of time or effort to the services than non-cherry pickers do. Our hypotheses are tested using data from a music streaming service in South Korea.

The paper is organized as follows. The next section reviews the literature relevant to cherry picking, price promotion, and switching behavior for online services. In Sect. 3, our model and dataset are explained. In Sect. 4, we examine the results of the hypotheses test, and the cherry picker prediction model will be discussed briefly. In Sect. 5, we discuss the limitations of our study and suggestions for future research.

2   Literature Review

In this section, we briefly discuss the previous research related to our study. They can be classified into three research topics. The first topic comprises price promotion strategies. The second topic includes customers' cherry-picking behavior mainly in offline circumstances. Lastly, we discuss online service switching behavior.

2.1   Price Promotion

Blattberg and Nesling [2] described the nature of sales promotion in their summary of common types of retailer discounts. One of such discount types is price reduction, which means that retailers temporarily decrease the prices of their products. The researchers examined four ways in which these promotions affect sales: (1) Branding-customers switch their purchases from other brands, (2) Stockpiling-the company's current consumers purchase higher quantities of the brand for inventory, (3) Purchase acceleration-current consumers accelerate their purchase of the good, and (4) Primary demand expansion-new consumers enter the market. Gazquez-Abad and Sanchez-Perez [4] showed that almost 47% of consumers can be considered deal prone, and sales promotions have a significant effect on customers' choice behavior.

The effect of price promotion seems positive however, Lal and Bell [8] found that promotions might have negative effect on a company's long-term profit. They found that frequent promotion increased the prevalence of cherry pickers while subsidizing loyal customers. One company's promotion can cause price wars with others. Van Heerde et al. [11] found that, although the price war initially entailed more shopping and spending, spending per visit ultimately dropped.

Many researchers have great interest in the topic of price promotion. Still, there is limited research on the price promotion in online subscription economy.

2.2   Cherry Picking Behavior

The definition of cherry picker by Fox and Hoch [3] is simple and clear. Cherry pickers opportunistically take the best and leave the rest in the contexts of both seller and buyer. On the buyer side, the term describes the behavior of buyers who are selective about which products or services they purchase at what locations and prices.

Talukdar et al. [10] studied extreme cherry picking behavior among consumers. They defined extreme cherry pickers as customers who seek deals and excessively avail themselves of deep discount offers, which generate negative profits for retailers. They found that the extreme cherry picking segment comprises about 2% of all shoppers.

To our best knowledge, most of the researches on cherry pickers is focused on offline rather than online environments, such as e-commerce.

2.3 Switching Behavior

Several researchers have tried to explain customers' online switching behavior using the Push-Pull-Mooring framework created by Bansal et al. [1]. They found that the attractiveness of other service pushes users to other services, while relative enjoyment, usefulness, and switching cost pull users [6, 14].

Keaveney and Parthasarathy [7] revealed that switchers demonstrate significantly lower service usage than continuers. While their results provided support for the relationship between service usage and continuance, their study considered free-to-use online services only.

Few studies have examined customers' online service switching behavior and most of them don't consider purchase, price, and price promotion.

3 Model

3.1 Hypotheses

Cherry picking users are users who can take the best among the services available to them. Under fixed subscription price for streaming service, the more users play streaming media, the more benefits users get. Therefore, our first hypothesis is as follows.

Hypothesis 1 (H1). Cherry picking users use online streaming services more actively than non-cherry picking users.

Hypothesis 1 is in contrast to the findings by Keaveney and Parthasarathy [7] regarding the relationship between service usage and continuance. Because their results relate to free services, we rather refer to the argument by Hackleman and Duker [5] in terms of the positive relationship between heavy usage and deal proneness. To test Hypothesis 1 in the music streaming service environment, two sub-hypotheses are addressed as follows.

Hypothesis 1a (H1a). The count of listening songs of cherry picking users is higher than that of non-cherry picking users.

Hypothesis 1b (H1b). The distinct number of songs played by cherry picking users is higher than that of non-cherry picking users.

Users investments, switching costs, or sunk costs are commonly indicated as penalties of switching services [6, 13]. With the previous research, we construct our second hypothesis and two sub-hypotheses as follows.

Hypothesis 2 (H2). Cherry picking users devote less investment to the service than non-cherry picking users.

Hypothesis 2a (H2a). The number of times cherry picking users "like" songs is smaller than that of non-cherry picking users.

Hypothesis 2b (H2b). The number of playlist-related behaviors among cherry picking users is smaller than that of non-cherry picking users.

3.2 Dataset

Our dataset consists of logged data of users of a music streaming service in South Korea over the course of eighteen months, from December 1, 2016 to June 30, 2018. We have three types of logs: purchase, listening behavior, and playlist management. In the case of the playlist management log, because the newest information is overwritten to the record, we only have the final status of users' playlists. We have every single transaction record for the other two types of log.

The provided price promotions are in effect only for the first three months. For example, users can use the service for a discounted price for three months at most. After three months, the price goes up to the regular price. Everyone can take the promotion only once. Even when a user signs out and signs up another account, he or she cannot receive the promotion again. Only the data of users who got the price promotion for their first-to-third purchase were used for our research. Out final dataset contains 834 users.

3.3 Variables

Our model comprises two categories of variables: usage and investment. The main purpose to subscribe to a music streaming service is to listen to music. We extract two variables representing users' usage: (1) avg_listen-, which is the daily average count of playing songs, and (2) avg_song-, the average distinct number of songs played per day. If a user's value of avg_listen is 15, it means that he or she played songs 15 times in total. It can be confusing to understand the variable avg_song. If a user's value of avg_song is 2, it means that he or she listened two unique songs every day. If this user subscribed for two months (60 days), his or her music pool consists of 120 songs. A high value of avg_listen but a low avg_song value indicates that the user listens to a few songs repeatedly.

When using the music streaming service, users can record their preference and create their own playlists. Marking their preference with the "like" button and managing his or her own playlists require time and effort. Thus, we utilize two variables for users' investment: (1) avg_like-, indicating the monthly average count of like flags, and (2) avg_playlist-, the monthly average count of playlist creation.

4 Results

4.1 Distribution

In the context of grocery shopping, Fox and Hoch [3] define customers' cherry picking behavior based on their shopping trip. If a customer has two or more shops in his or her shopping trip, he or she is considered a cherry picker. We define cherry picking users as those who only purchase good at a discounted price, not the regular price. This means that they quit using a service right after or during the promotion period.

In other words, cherry picking users subscribe to the music streaming service for at most three months. Non-cherry picking users subscribe to the service for at least five months. We exclude the users who subscribed to the service for exactly four months. Because there are chances that, even though they were cherry pickers, they forgot to quit before the price rose. Following these criteria, we have 345 cherry picking users and 489 non-cherry picking users within our sample.

Our data shows that 41.37% of users are cherry picking users. This ratio is much higher than the previous findings of Talukdar, et al. [10]. Their results are given within the context of grocery shopping. The high ratio can be examined by the fact that it is much easier for customers to switch from one service to another.

4.2 Hypotheses Test

To test our hypotheses, we apply an independent two-sample t-test with unequal variances. In the context of user usage, we include two variables: *avg_listen* and *avg_song*. There was a significant difference in *avg_listen* between the cherry pickers (M = 24.42; SD = 75.61) and non-cherry pickers (M = 16.86; SD = 46.89). The 95% confidence intervals for two means are shown on the left side of Fig. 1 (t = −2.45, p-value = 0.01, H1a was supported). For avg_song, there was a significant difference between that of the cherry pickers (M = 4.09; SD = 9.00) and non-cherry pickers (M = 1.68; SD = 2.25). The 95% confidence intervals for two means are shown on the right side of Fig. 1 (t = −5.98, p-value = 0.00, H1b was supported). Thus, Hypothesis 1 was supported.

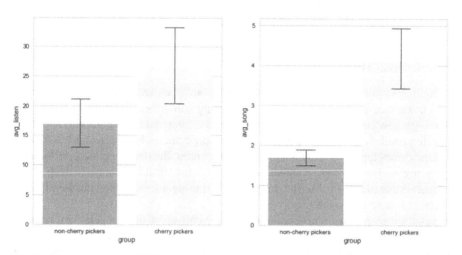

Fig. 1. Observed mean and 95% confidence interval of *avg_listen* (left) and *avg_song* (right)

In terms of the investment context, we have two variables: *avg_like* and *avg_playlist*. There was a significant difference in *avg_like* between the cherry pickers (M = 3.78;

SD = 18.62) and non-cherry pickers (M = 0.56; SD = 2.22). The 95% confidence intervals for two means are shown on the left side of Fig. 2 (t = −3.95, p-value = 0.00, H2a was not supported). For *avg_playlist*, there was a significant difference between the cherry pickers (M = 1.06; SD = 3.13) and non-cherry pickers (M = 0.27; SD = 0.68). The 95% confidence intervals for two means are shown on the right side of Fig. 2 (t = −5.66, p-value = 0.00, H2b was not supported). Interestingly, we found the exact opposite result from our expectation. Following the study by Ray et al. [9], we found that users' technical abilities outweigh user-related switching costs. In short, users with technical self-efficacy can easily re-generate their playlist and "like" markers.

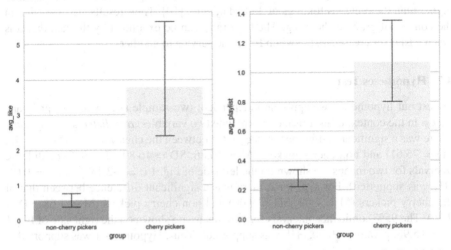

Fig. 2. Observed mean and 95% confidence interval of *avg_like* (left) and *avg_playlist* (right)

4.3 Prediction Model

To acquire new subscribers through price promotions, companies should expend money. It is known that the cost of retaining an existing subscriber is generally lower than acquiring a new customer. Thus, it is critical for the companies to predict cherry pickers when they conduct marketing strategies like price promotions. Early research has focused on users' demographic information to predict user churn. But demographical information is sensitive and using demographical data renders the resulting churn analysis at the customer rather than the subscriber level [12]. In this regard, we exploit the use of users' behavior extracted from the logs.

Based on interviews with marketing representatives of the cooperative company, nine variables potentially affecting a user's churn are identified. The description of the variables is shown in Table 1. Unlike the hypotheses test, we narrow the observation scope of users to 24 h after their initial payment.

We build four prediction models using decision tree (DT), logistic regression (LR), random forest (RF), and support vector machine (SVM) respectively. We evaluate the performance of each model by F1-score, the harmonic mean of precision and recall

Table 1. Summary of explanatory variables

Variable	Description
count_listen	The count of playing songs within 24 h after a user's initial payment
count_song	The distinct number of songs played within 24 h after a user's initial payment
chart_ratio	The ratio of songs in popular charts to songs played within 24 h after a user's initial payment
like_ratio	The ratio of songs "liked" to songs played within 24 h after a user's initial payment
nMins_to_play	Time elapsed for a user's first playing a song after his or her initial payment (in minutes)
use_playlist	Whether a user generates his or her own playlist
nHours_to_playlist	Time elapsed for a user's first generation of a playlist after his or her initial payment (in hours)
use_like	Whether a user "likes" a song at least once
nHours_to_like	Time elapsed for a user's first "like" after his or her initial payment (in minutes)

(DT: 0.64; LR: 0.66; RF: 0.72; SVM: 0.7). The F1-scores are averaged after 5-fold cross validation and all models outperforms random guess (0.5).

5 Conclusion

In this paper, we investigate the characteristics of cherry pickers in subscription-based online services. Online is a good place for customers to pick cherries, and cherry picking behavior online might be different from that offline. Our results show that cherry pickers use online music streaming services more actively and apply their investment to the service more readily than non-cherry pickers. We believe that our findings help to understand customers' cherry picking behavior online.

Our study has several limitations that provide further opportunities. First, we only consider a music streaming service. Other types of subscription services should be considered. Second, we don't approach to a prediction model with great performance. If cherry pickers can be detected or predicted during an early period, much more implications for both theoretical and practical would be available. Furthermore, any causal relationship is not revealed.

As Talukdar et al. [10] stated, users' cherry-picking behavior can generate negative profits for service providers. To resolve this problem, service providers must find an effective way to entice even cherry pickers to commit to their services. This is fruitful area for possible future research.

References

1. Bansal, H.S., Taylor, S.F., St. James, Y.: Migrating to new service providers: toward a unifying framework of consumers' switching behaviors. J. Acad. Mark. Sci. **33**(1), 96–115 (2005)
2. Blattberg, R.C., Neslin, S.A.: Sales promotion. Oxford Handbook, Englewood Cliffs (1990)
3. Fox, E.J., Hoch, S.J.: Cherry-picking. J. Mark. **69**(1), 46–62 (2005)
4. Gazquez-Abad, J.C., Sanchez-Perez, M.: Characterising the deal-proneness of consumers by analysis of price sensitivity and brand loyalty: an analysis in the retail environment. Int. Rev. Retail: Distrib. Consum. Res. **19**(1), 1–28 (2009)
5. Hackleman, E.C., Duker, J.M.: Deal proneness and heavy usage: Merging two market segmentation criteria. J. Acad. Mark. Sci. **8**(4), 332–344 (1980)
6. Hsieh, J.K., Hsieh, Y.C., Chiu, H.C., Feng, Y.C.: Post-adoption switching behavior for online service substitutes: a perspective of the push–pull–mooring framework. Comput. Hum. Behav. **28**(5), 1912–1920 (2012)
7. Keaveney, S.M., Parthasarathy, M.: Customer switching behavior in online services: an exploratory study of the role of selected attitudinal, behavioral, and demographic factors. J. Acad. Mark. Sci. **29**(4), 374–390 (2001)
8. Lal, R., Bell, D.E.: The impact of frequent shopper programs in grocery retailing. Quantitative Mark. Econ. **1**(2), 179–202 (2003). https://doi.org/10.1023/A:1024682529912
9. Ray, S., Kim, S.S., Morris, J.G.: Research note—online users' switching costs: their nature and formation. Inf. Syst. Res. **23**(1), 197–213 (2012)
10. Talukdar, D., Gauri, D.K., Grewal, D.: An empirical analysis of the extreme cherry picking behavior of consumers in the frequently purchased goods market. J. Retail. **86**(4), 336–354 (2010)
11. Van Heerde, H.J., Gijsbrechts, E., Pauwels, K.: Winners and losers in a major price war. J. Mark. Res. **45**(5), 499–518 (2008)
12. Wei, C.P., Chiu, I.T.: Turning telecommunications call details to churn prediction: a data mining approach. Expert Syst. Appl. **23**(2), 103–112 (2002)
13. Yang, Z., Peterson, R.T.: Customer perceived value, satisfaction, and loyalty: the role of switching costs. Psychol. Mark. **21**(10), 799–822 (2004)
14. Zhang, K.Z., Cheung, C.M., Lee, M.K.: Online service switching behavior: the case of blog service providers. J. Electron. Commerce Res. **13**(3), 184–197 (2012)

The Value of Free Content on Social Media: Evidence from Equity Research Platforms

Tianyou Hu[(✉)] [iD], Arvind Tripathi [iD], and Henk Berkman

The University of Auckland, 12 Grafton Road Sir Owen G Glenn Building, Auckland 1010, New Zealand
{t.hu,a.tripathi,h.berkman}@auckland.ac.nz

Abstract. The effect of social media sentiments on stock market returns is well-established. However, the quality of content and expertise of content creators vary on social media platforms, and the stocks vary in characteristics. In this research, we examine the effect of sentiment expressed in free content from a social media platform on stock abnormal returns. We also examine the moderating effect of the market capitalisation of stocks on the strength of this relationship. Using data collected from a well-known equity research platform, we demonstrate that the size of the market cap plays an important role in this relationship. The smaller the market cap, the higher the predicting power of the social media sentiment on stock abnormal returns. Considering different holding periods from 1 month to 1 year, we show that sentiments from social media have a long wear in effect on stock abnormal returns. Our results shed light on the importance of market cap and holding period when studying the effect of social media sentiments on stock market returns.

Keywords: Social media · Stock returns · Market cap

1 Introduction

Retail Investors have always relied on public and easily accessible information for investment advice. Starting from the newspaper in the old days, to TV shows and finally, on digital media, there has always been a demand for financial investment information. With the proliferation of the Internet and ubiquitous devices, users now consume financial news and other market information from platforms like Yahoo Finance, Google Finance and Bloomberg. Although news media reaches a wide range of audience, it can't compete with social media platforms in terms of reach that thrive on user participation because users are able to share their views instantaneously. Further, powered by the wisdom of crowds, specialised social media forums often provide more valuable information in financial markets compared to news aggregators such as Google Finance (Hu, Srinivasan and Tripathi 2018).

There are many types of social media platforms based on content and contributors. At one end, we have social media platforms such as Twitter, where users are allowed to post/share anything with other users. However, on these platforms, often we observe a

© Springer Nature Switzerland AG 2020
K. R. Lang et al. (Eds.): WeB 2019, LNBIP 403, pp. 123–130, 2020.
https://doi.org/10.1007/978-3-030-67781-7_12

mix of experts and novice users posting similar or conflicting information, often leading to contradictory recommendations. These challenges have given rise to specialised social media forums such as StockTwits, RagingBull and many others. While these specialised forums are better than general social media forums in terms of the overall content, focus and quality of the information, contributors of these platforms can be anyone without any finance background. Thus, the question of information quality remains unaddressed. To combat these issues and meet the market demand for quality investment information, some equity research platforms were started. One such example is Seeking Alpha (SA, https://seekingalpha.com/), which was founded in 2004. According to SA website, their editors "curate content from a network of stock analysts, traders, economists, academics, financial advisors and industry experts". Seeking Alpha (SA) is an equity research platform with social media features. But the content published on SA is carefully analysed by editors for quality. Authors who write articles on SA get paid by the platform.

SA provides users with the opportunity to tap into the Wisdom of Crowds, where users can get information from many individuals which often turns out to be better than a few experts (Chen et al. 2014). However, we are yet to understand how different types of content and content creators can successfully predict market returns. We need to understand why experts create this valuable content, which is also available for free consumption in certain cases. Several factors may play a role: (1) Significant utility can be drawn from the attention and recognition from posting ideas and being confirmed by the market. (2) SA contributors are paid $10 per 1000-page views and can earn at least $500 more if the articles is considered to be of high quality. (3) SA allows users to directly chat with each other and provide timely and publicly visible feedback, which can correct bad articles. (4) It may have some price impact if the SA users read an article and trade based on the idea (Chen et al. 2014).

A few studies have examined content created on Seeking Alpha and its effect on market returns. Researchers have shown that views disclosed in both articles and commentaries predict future stock abnormal returns and earnings surprise (Chen et al. 2014). While Seeking Alpha publishes both free and paid content, the value of content and advice given via free articles is yet to be studied. Further, it is not clear if the effect of the advice/sentiment expressed in these free articles varies based on market caps of the predicted stocks. To address these issues, this research aims to answer the following research questions: Do sentiments expressed in free articles on SA predict market returns? Is this relationship impacted by the market cap of the stocks predicted? How does this effect change overtime?

This paper is organised as follows. Section 2 presents the literature review. Section 3 describes the data collection. Section 4 details the methodology. Section 5 presents the results. Section 6 sheds light on the conclusion.

2 Literature Review

In recent years, researchers have examined the effect of sentiments expressed on social media on financial markets. However, these studies have found conflicting results, which we discuss next.

Bollen, Mao and Zeng (2011) find that aggregated sentiment resulted from sentiment analysis of daily tweets on Twitter can help predict movements in the Dow Jones Index.

Mao, Wei, Wang and Liu (2012) show that the daily amount of tweets that mention S&P 500 stocks is significantly correlated with the levels, changes, and absolute changes in the S&P 500 Index.

Stock discussion forums are different than general social media platforms, such as Twitter, in terms of their influence on financial markets. For example, a recent study examining sentiments on stock discussion forums (Tumarkin and Whitelaw, 2001) has demonstrated that on days of abnormal board activity, message board opinions and stock returns are correlated, but there is no evidence that opinion anticipates future returns. Antweiler and Frank (2004) find that significant negative returns follow a higher discussion forum posting volume. However, the economic impact will be small. Das and Chen (2007) present that the aggregated high-tech sector sentiment is correlated with high-tech sector index returns, but not for individual stocks. Leung and Ton (2015) show that the number of messages and message sentiment on a stock discussion form HotCopper, have an effect on the contemporaneous stock returns.

While some studies establish the effect of stock discussion forums on social media on market movement, there are others who claim the opposite. For example, Kim and Kim (2014) show that there is no evidence that investor sentiment predicts future stock returns. Chen et al. (2014) find that views expressed by experts on equity research platforms and users' comments predict future stock abnormal returns and earnings surprise examining data from SA. While there is plenty of literature investigating the effect of sentiments, it is not clear if these discussions on social media and discussion forums only focus on large-cap stocks or alternatively, whether the size of market cap mediates the effect of sentiments on stock returns. We attempt to address this question in this research.

Recent studies have demonstrated that large-cap stocks receive a higher traditional media coverage, e.g. newspaper and TV (Tetlock, 2007; Tetlock, Saar-Tsechansky and MacSkassy, 2008; Fang & Peress, 2009; Engelberg and Parsons, 2011) compared to small-cap stocks. Since small-cap stocks often suffer from a lack of coverage by media and analysts (Tetlock, 2007), the gap is filled by social media and discussion forums where users share their opinions and information with other investors (Ahmed et al. 2003). In addition to these studies, it is shown that there is a stronger positive relationship between Mad Money (a TV show) buy recommendation and abnormal returns and trading volumes for small-cap stocks compared to large-cap ones (Keasler and McNeil 2010). Another study shows that higher social media activities for small stocks are linked with optimistic sentiment (Zhang et al. 2012).

In this study, we hypothesise that the market cap will inversely affect the impact of social media sentiment on stock market abnormal returns. In other words, sentiments on social media will be more effective for predicting market returns for small market cap stocks compared to large caps. Tumarkin and Whitelaw (2001) show that company and sector experts tend to disclose value-relevant information on the Internet, potentially after building a long position on the stocks themselves. Boehme, Danielsen and Sorescu (2009) show that due to the high cost of the short sale, online investors are more willing to gather information on stocks they are considering starting a long position instead of spreading false information to generate profit.

While current literature investigates the effect of sentiments on social media on market movement, such as abnormal returns, volatility and trading volume, but do not examine if the market cap of the stocks plays any role in this relationship.

3 Data

Articles submitted to SA are reviewed by editors for clarity, consistency and impact. The purpose of this review process is to increase readability without changing the authors' original point of view. Authors are required to disclose their positions (long, short or no position) in stocks they write about. Authors are paid based on the number of page views and earn extra if their articles are selected to be of particularly high quality by editors.

SA publishes two types of articles: Free and Pro. To read pro articles, users have to pay a subscription fee. As expected, free articles attract more views than pro articles. In this research, we have only considered free articles, as free articles are more visible to all online investors. Further, articles on SA discuss a wide range of stocks, from most popular to barely known, from large-cap to small-cap, etc. We have only considered stocks which have been covered by seeking alpha free articles, which is free content.

Each article is tagged with one or more stock tickers, and articles have clear disclosure positions. In this research, we only consider articles that have directional self-disclosure positions (long or short) from the authors. An article has listed the sentiments for one or more tickers. We consider the disclosure positions (sentiments) for all these tickers. SA has only enabled disclosure from 2015. To answer the above research question, we have collected free articles from SA span from 2015 to 2018. After these procedures, our dataset boils down to 1184 tickers and 24420 free articles that have directional disclosure positions.

Stocks are segmented (Table 1) into three groups based on the size of market cap as recommended by popular investment news platform, Investopedia. Market cap bigger than 10 billion is categorised as large-cap, smaller than 2 billion as small-cap, and rest as mid-cap.

Table 1. Number of tickers in different market cap categories

Market cap	Large	Mid	Small
Number of tickers	167	261	756

4 Methodology

First, we compute the standardised bullishness index $Bullishness_{i,t}$ (Antweiler and Frank, 2004) for stock i on day t. It can be calculated as follows:

$$Bullishness_{i,t} = \frac{M_{i,t}^{Bullish} - M_{i,t}^{Bearish}}{M_{i,t}} * ln\left(1 + M_{i,t}\right) \tag{1}$$

$M_{i,t}^{Bullish}$ is the number of articles with "Bullish" sentiment, $M_{i,t}^{Bearish}$ is the number of articles with "Bearish" sentiment. And $M_{i,t} = M_{i,t}^{Bullish} + M_{i,t}^{Bearish}$ is the total number of relevant articles. Note that these sentiments are explicitly expressed by the authors on SA.

Previous research has used raw return, which is the natural logarithm of the last holding day's adjusted close price divided the adjusted close price on first day (Kim and Kim, 2014; Leung and Ton, 2015) and abnormal returns and earnings surprise (Chen et al. 2014).

To examine the effect of explicit sentiments expressed by authors in free articles on stock abnormal returns, we propose the following model.

$$
\begin{aligned}
AReturn_{i,t\_holding\_perid} = {} & \alpha + \beta_1 MarketCapIsLarge + \beta_2 MarketCapIsMid \\
& + \beta_3 Bullishness_{i,t} + \beta_4 MarketCapIsLarge * Bullishness_{i,t} \\
& + \beta_5 MarketCapIsMid * Bullishness_{i,t} + \varepsilon
\end{aligned}
\tag{2}
$$

Abnormal returns are the company's raw returns minus the return of a value-weighted portfolio with similar size/book-to-market/past return-characteristics using the following holding periods: one month, three months, six months and one year. t is the date when the article is published or the first trading day following the publication date if it is published on a non-trading date (Chen et al. 2014). $AReturn_{i,t\_holding\_period}$ is the buy and hold abnormal return for varied holding periods starting from the next trading day after time t. In this research, we categorise stocks into large, mid and small caps (Table 1). $MarketCapIsLarge$ and $MarketCapIsMid$ are dummy variables and are 1 for large and midcap stocks and 0 otherwise, respectively. The dummy variable value allocation of each market cap category is shown in Table 2. We use the interaction term of $MarketCapIsLarge * Bullishness_{i,t}$ and $MarketCapIsMid * Bullishness_{i,t}$ to demonstrate how the size of Market Cap influences the relationship between the Bullishness index and the dependent variable which is the abnormal returns.

Table 2. Market cap categories and dummy variable value allocation

Market cap	Large	Mid	Small
MarketCapIsLarge	1	0	0
MarketCapIsMid	0	1	0

5   Results

Table 3 shows the summary of the results of our proposed model in Eq. (2). *t-statistics* are reported in parentheses.

In this model, we test the effect of sentiment on social media via free articles on Seeking Alpha, on stock abnormal returns. Sentiments of social media are measured by Bullishness index. Using two dummy variables, *MarketCapIsLarge* and

Table 3. Results

	1 month	3 months	6 months	12 months
MarketCapIsLarge	0.020258	0.046787	0.1431709	0.52321
	(3.288)	(3.935)	(6.700)	(9.436)
MarketCapIsMid	0.020395	0.051235	0.1341671	0.48064
	(3.437)	(4.474)	(6.519)	(9.000)
$Bullishness_{i,t}$	0.021072	0.029265	0.0684869	0.19922
	(5.587)	(4.021)	(5.236)	(5.870)
$Bullishness_{i,t}$ * MarketCapIsLarge	−0.018073	−0.022705	−0.056521	−0.19047
	(−2.925)	(−1.904)	(−2.637)	(−3.425)
$Bullishness_{i,t}$ * MarketCapIsMid	−0.012303	−0.002851	−0.000427	−0.08002
	(−2.065)	(−0.248)	(−0.021)	(−1.492)

MarketCapIsMid, we have also attempted to examine if the effect of social media sentiment on stock returns vary for large, mid or small-cap stocks with different holding periods. If MarketCapIsLarge and MarketCapIsMid are both 0 (small-cap stocks), the aggregated coefficient estimates of $Bullishness_{i,t}$ are always positive and significant for all for the different holding periods. For instance, if the Bullishness increase by 1, then the abnormal return will increase by 2.1072% for small-cap stocks with one month holding period.

With the coefficient estimates of $Bullishness_{i,t}$ * MarketCapIsLarge and $Bullishness_{i,t}$ * MarketCapIsMid being negative for all the holding periods, the effect of expert sentiments measured via Bullishness Index is strongest for small-cap stock and weaker for large and medium cap stock for all the different holding periods. The result so far shows that aggregated sentiment from social media does predict subsequent stock abnormal returns.

Overall, we observe that the predicting power of social media sentiment decreases with the increase in market cap. Prior research (Hu and Tripathi 2018) has shown that investments made based on advice offered in free articles on SA result into positive returns. However, it wasn't clear if the market cap of stocks plays any role in moderating the advice offered in free articles published on Seeking Alpha.

The coefficient estimates of the $Bullishness_{i,t}$ of small-cap stocks tend to increase with the length of the holding period. This result is a bit different from the observations in the financial analyst literature, which most of the abnormal performance surrounds the date of the recommendation change. The reason can be that even with SA's increasing popularity, financial analysts still get a lot more attention among investors. As a result, analysts' opinion is reflected in the market at a faster pace. In comparison, social media get a bit less attention among investors, and free content is reflected in the market at a slower pace. With the proliferation of social media and the increasing value of them, one may speculate that the abnormal performance will accrue around the initial publication date of free articles from social media.

6 Conclusion

Advancement in Web 2.0 technologies has led to the proliferation of social media platforms as dominant channels in online markets. Since social media platforms vary in terms of quality of content and expertise of content creators, it is important to understand and differentiate the value content created on these platforms in a variety of online markets. The question of content quality becomes more important in financial markets due to increased risks and rewards. Prior studies (Hu and Tripathi 2018) have investigated the impact of sentiments expressed on social media platforms on market returns and have shown that free advice offered by experts on Seeking Alpha is highly valuable. Our study adds to this finding. Our results show that while sentiments expressed by experts in free articles on Seeking Alpha predict market returns, the market cap of the stocks plays a role in moderating this relationship. In other words, free articles are more valuable for small-cap stocks compared to large-cap stocks. It is also shown that the abnormal performance predicted by SA articles happen at a slow pace. However, with the advancement of social media, it can be expected that this kind of abnormal performance will move towards the publication date of articles.

Overall, this study shows how sentiments expressed on social media can affect the returns of stocks with different market caps. This research answers the question of whether the size of the market cap influences the predicting power of social media sentiments.

In summary, we find that social media sentiments have a larger predicting power over the stocks with smaller market caps.

References

Ahmed, A.S., Schneible Jr., R.A., Stevens, D.E.: An empirical analysis of the effects of online trading on stock price and trading volume reactions to earnings announcements. Contemp Account Res. **20**, 413–439 (2003)

Antweiler, W., Frank, M.Z.: Is all that talk just noise? The information content of Internet stock message boards. J. Finance **59**, 1259–1294 (2004). https://doi.org/10.1111/j.1540-6261.2004.00662.x

Boehme, R.D., Danielsen, B.R., Sorescu, S.M.: Short-sale constraints, differences of opinion, and overvaluation. J. Financ. Quant. Anal. **41**, 455 (2009). https://doi.org/10.1017/S0022109000002143

Bollen, J., Mao, H., Zeng, X.: Twitter mood predicts the stock market. J. Comput. Sci. **2**, 1–8 (2011). https://doi.org/10.1016/j.jocs.2010.12.007

Chen, H., De, P., Hu, Y.Y. (Jeffrey), Hwang, B-HB-H: Wisdom of crowds: the value of stock opinions transmitted through social media. Rev. Financ. Stud. **27**, hhu001 (2014). https://doi.org/10.1093/rfs/hhu001

Das, S.R., Chen, M.Y.: Yahoo! for Amazon: sentiment Extraction from Small Talk on the Web. Manage. Sci. **53**, 1375–1388 (2007)

Engelberg, J.E., Parsons, C.A.: The causal impact of media in financial markets. J. Finance **66**, 67–97 (2011). https://doi.org/10.1111/j.1540-6261.2010.01626.x

Fang, L., Peress, J.: Media coverage and the cross-section of stock returns. J. Finance **64**, 2023–2052 (2009). https://doi.org/10.1111/j.1540-6261.2009.01493.x

Hu, T., Tripathi, A.: Is there a free lunch? examining the value of free content on equity review platforms. In: Cho, W., Fan, M., Shaw, M., Yoo, B., Zhang, H. (eds) Digital Transformation: Challenges and Opportunities. WEB 2017. Lecture Notes in Business Information Processing, vol. 328. Springer, Cham (2017)

Hu, T., Srinivasan, A., Tripathi, A.: Analysing sentiments from social media forums: the case of financial markets. In: Proceedings of Mediterranean Conference of Information Systems (MCIS), 2018 (2018)

Keasler, T.R., McNeil, C.R.: Mad Money stock recommendations: market reaction and performance. J. Econ. Financ. **34**, 1–22 (2010). https://doi.org/10.1007/s12197-008-9033-7

Kim, S.-H., Kim, D.: Investor sentiment from internet message postings and the predictability of stock returns. J. Econ. Behav. Organ. **107**, 708–729 (2014). https://doi.org/10.1016/j.jebo.2014.04.015

Leung, H., Ton, T.: The impact of internet stock message boards on cross-sectional returns of small-capitalisation stocks. J. Bank Financ. **55**, 37–55 (2015). https://doi.org/10.1016/j.jbankfin.2015.01.009

Mao, Y., Wei, W., Wang, B., Liu, B.: Correlating S&P 500 stocks with Twitter data. In: Proceedings of the First ACM International Workshop on Hot Topics on Interdisciplinary Social Networks Research - HotSocial 2012. ACM Press, New York, New York, USA, pp. 69–72 (2012)

Tetlock, P.C.: Giving content to investor sentiment: the role of media in the stock market. J. Finance **62**, 1139–1168 (2007). https://doi.org/10.1111/j.1540-6261.2007.01232.x

Tetlock, P.C., Saar-Tsechansky, M., MacSkassy, S.: More than words: quantifying language to measure firms' fundamentals. J. Finance. **63**, 1437–1467 (2008). https://doi.org/10.1111/j.1540-6261.2008.01362.x

Tumarkin, R., Whitelaw, R.F.: News or noise? internet postings and stock prices. Financ. Anal. J. **57**, 41–51 (2001). https://doi.org/10.2469/faj.v57.n3.2449

Zhang, Y., Swanson, P.E., Prombutr, W.: Measuring effects on stock returns of sentiment indexes created from stock message boards. J. Financ. Res. **35**, 79–114 (2012). https://doi.org/10.1111/j.1475-6803.2011.01310.x

Managing e-Business Projects and Processes

Managing Cloud Computing Across the Product Lifecycle: Development of a Conceptual Model

Timo Puschkasch[(✉)] and David Wagner

Munich Business School, Elsenheimerstraße 61, 80687 Munich, Germany
{timo.puschkasch,david.wagner}@munich-business-school.de

Abstract. Cloud computing has become an important part of IT infrastructure for both small companies and large enterprises over the last years. More organizations than ever consider it an enabler for their efforts in making IT more agile, reducing operating costs and gaining access to new technologies that will give them an edge over their competitors. In this paper we develop a conceptual model to explain the benefits of using cloud computing for delivering a digital product across several stages of its product lifecycle, emphasizing the importance of the four delivery models: public, private, community and hybrid cloud. While this distinction is theoretically novel and helps IS scholars to gain a more nuanced understanding of cloud computing in the enterprise, it provides practitioners with a decision frame to select the most effective mode of cloud computing given the specific lifecycle stage of their digital product. Ultimately, we discuss limitations of this approach and provide directions for future research.

Keywords: Digital business models · Cloud computing · Product lifecycle · Agility · Delivery models

1 Introduction

Cloud computing describes a way for organizations to obtain "ubiquitous, convenient, on-demand network access to a shared pool of configurable computing resources" [1]. By the end of 2019, an additional 17% of enterprises are planning to migrate at least partly to Cloud computing according to IDG [2], one of the largest IT media and research companies worldwide, and will therefore face the decision as to which kind of cloud environment is right for their product. According to research company Gartner, by 2020 a "no-cloud policy" will not be an option anymore if companies want to stay competitive [3].

Cloud computing is considered a – and possibly the - strategic driver of digital product development [4, 5]. Traditionally, the product lifecycle model was used to explain the requirements for successful product development given the specific lifecycle stage of a product [6]. Digital products are no exception [7–9].

While extensive research has been conducted into the technological aspects and – to a lesser extent – business issues of cloud computing, conceptual research papers are still significantly underrepresented [10–12]. In a recent review of the cloud computing literature, Senyo et al. [13] find that only about 9% of the 285 articles under scrutiny were

© Springer Nature Switzerland AG 2020
K. R. Lang et al. (Eds.): WeB 2019, LNBIP 403, pp. 133–142, 2020.
https://doi.org/10.1007/978-3-030-67781-7_13

an attempt to conceptualize the phenomenon, while the vast majority of papers (235) were classified as atheoretical, meaning that no theoretical frame was provided. Attempting to address this shortcoming, the objective of this paper is to create a conceptual model, mapping the benefits of cloud computing to the product lifecycle stages of a digital product. We develop this conceptual model by using an established theory, i.e. the product lifecycle model, to explain a new phenomenon, i.e. cloud computing, following one of the approaches put forward by Yadav [14]. While this distinction is theoretically novel and helps IS scholars to gain a more nuanced understanding of cloud computing in the enterprise, it provides IT managers responsible for cloud computing with a decision frame to select the most effective mode of cloud computing given the specific lifecycle stage of their digital product.

The paper is structured as follows: we will first introduce the theoretical background of cloud computing and the product lifecycle model. Next, we will develop our conceptual model by analyzing the challenges of each lifecycle stage and linking them with the benefits cloud computing can provide. We will then discuss the model and its limitations before we conclude with recommendations for future research.

2   Cloud Computing

Cloud computing as a technological trend allows organizations to quickly and easily obtain IT resources over a network connection, for example through the public internet [1]. As such it is a strategic driver behind developing digital products and enabling digitization of existing products [4, 5]. Extensive research has been conducted about the phenomenon of cloud computing, e.g. on the benefits of cloud computing for organizational agility [11] or even the impact of cloud computing adoption on stock prices of large enterprises [15].

2.1   Deployment Modes of Cloud Computing

One common definition of cloud computing is the NIST standard, developed by Mell and Grance [1]. In the standard, a differentiation between service models and deployment models is made. The term service models uses a simple taxonomy to distinguish between various degrees of integration. Deployment models on the other hand describe the degree of isolation of a cloud offering and, therefore, the amount of control the consumer can exercise over it. For the purpose of this paper, the differentiation between service models will be omitted when discussing the benefits of Cloud computing as the difference across the product lifecycle is mainly driven by the deployment model used. The impact of different service models is slim by comparison. The deployment models can be defined as follows [1]:

- *Private Cloud*: The cloud infrastructure is provisioned for exclusive use by a single consumer
- *Community Cloud*: A community cloud is a special form of a Private Cloud where several consumers, for example companies within a large enterprise, join to create a Cloud environment that is exclusive to members of this group

- *Public Cloud*: The cloud infrastructure is provisioned for open use by the general public
- *Hybrid Cloud*: The cloud infrastructure is a composition of two or more distinct cloud infrastructures (private, community, or public) that remain as unique entities, but are bound together by standardized or proprietary technology that enables data and application portability (e.g., cloud bursting for load balancing between clouds)

2.2 Benefits of Cloud Computing

Marston et al. [12] have compiled a list of benefits that can be realized from a business perspective by using cloud computing:

- *Lower Cost of Entry*: Depending on the deployment model, cloud computing requires little to no upfront investment in infrastructure, making it easier for enterprises and new competitors to enter the market
- *Faster Provisioning Times*: Cloud computing can allow quicker access to IT resources, therefore reducing the time needed to provision new services
- *Lower Barrier to Innovation*: Due to access to new technologies and faster provisioning times cloud computing can be used to develop and experiment quickly
- *High Scalability of Resources*: Cloud computing is designed to scale both up and down quickly, allowing to react to higher demand and at the same time freeing resources if they are no longer required
- *Access to Specialized Technology*: Depending on the deployment model, cloud computing can provide access to new technologies and skills that an organization would not be able to provide internally.

Additionally, Litoiu et al. [16] have observed that cloud computing can be optimized for specific workloads and the ability to optimize depends on the deployment model used. For example, according to Litoiu et al. [16], public cloud computing is often made up of "a layered architecture, where a feedback loop has only a limited view of resources and goals." Hence, the optimization techniques commonly used in private cloud computing, where full transparency of resources can be achieved, are not fully applicable here. This transparency given in private cloud environments in addition to the ability to make individual changes to the underlying architecture which would not be possible in public cloud environments allows for better optimization towards the specific workload.

The benefits described above can be mapped to the deployment model based on their degree of applicability. Generally, all deployment models can achieve faster provisioning times compared to traditional IT infrastructure. However, for other benefits, differences exist. If we take lower cost of entry as an example, Private Cloud would provide only a slight benefit regarding the lower cost of entry, as investments still need to be made to establish the cloud environment, similar to traditional IT infrastructure. On the other hand, in a Public Cloud environment, these resources are pooled by a cloud provider and can be accessed on-demand for a usage-based fee, leading to a high benefit for this deployment model. In community cloud environments, the pool of shared resources is limited due to the reduced amount of organizations sharing the cloud environment, resulting in a medium benefit. The same benefit is applied to hybrid cloud environments, as the cost of entry is low for public resources, but high for private resources, which are both part of the hybrid environment. The remaining benefits will not be discussed in detail; however, we followed a similar logic to derive the degree of applicability for each benefit. A summary of our analysis is displayed in Table 1.

Table 1. Benefits of Cloud computing depending on deployment model

Benefit of cloud computing	Degree of applicability to deployment models			
	Private cloud	Community cloud	Public cloud	Hybrid cloud
Lower cost of entry	Low	Medium	High	Medium
Faster provisioning times	High	High	High	High
Lower barrier to innovation	Medium	Medium	High	High
High scalability of resources	Medium	Medium	High	High
Access to specialized technology	Low	Medium	High	High
Ability to optimize for specific workloads	High	Medium	Low	High

While other benefits of Cloud computing have been described, we consider these to be the most relevant for the purpose of mapping them to the product lifecycle.

3 The Product Lifecycle

The product lifecycle is a tool first introduced by Vernon [6], which has quickly been adopted as one of the fundamental instruments in economics [17, 18]. According to Cox [17], the development of any product can be classified into four stages:

1. *Introduction Stage*: Initially, an idea is validated and at some point, offered to a specialized market or the broad market, thereby being introduced
2. *Growth Stage*: If a product is sustainable enough in the introduction stage, it will gain recognition in the market and begin to grow
3. *Maturity or Stabilization Stage*: After a certain time of growth, all products will eventually reach a point where sales incline stops. This phase can be triggered early by competitors entering the market, leaving less market potential to the original product
4. *Decline Stage*: Once a product has reached maturity for some time, sales will inevitably decline, and revenue will begin to fall

It is worth noting at this point that a product may fail in any of these stages and will then move directly to the last step, cancellation, without following through with the remaining steps of the PLC [17]. A visual representation of the four stages can be seen in Fig. 1.

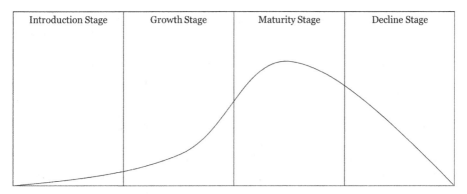

| Introduction Stage | Growth Stage | Maturity Stage | Decline Stage |

Fig. 1. The product lifecycle, adapted from Levitt [19]

Digital products are different from traditional, manufactured products in many ways. For example, the environment is more volatile and competitive, leading to an uncertainty in market and product development [20]. However, existing research has successfully applied the traditional product lifecycle model to digital products. One good example is the application to the field of mobile apps [8], as these apps make up a large market with about 178 billion downloads in 2017 [21] and therefore provide a large basis for analysis. Furthermore, researchers have tied the product lifecycle model to digital products in a variety of ways [7, 9]. Therefore, we conclude that the stages laid out in the original product lifecycle model [6] equally apply to digital products.

4   Using Cloud Computing for Product Lifecycle Management

4.1   Typical Challenges Associated with Lifecycle Stages

Using the product lifecycle model, distinct challenges of each stage can be defined [17, 19]. The Introduction stage is mainly driven by an uncertainty of market success for the new product and a high up-front investment, which is required to develop and launch a product. During the Growth phase, if the product was successful during its introduction, the risk of competition entering the market to compete for market share will arise. Once the Maturity phase has been reached, the market will begin to be saturated, therefore slowing the growth rate of the product. Given the existing competition on the market by this point, this will lead to increased price pressure, as companies start to compete on price to gain additional market share. Lastly, in the decline phase, market demand will shrink, leading to lower utilization of the infrastructure and resources used for the product.

4.2 Using Cloud Computing to Address Lifecycle Challenges

It can be assumed that the change in challenges across the product lifecycle stages will also require a different setup of the underlying resources and infrastructure [22], of which cloud computing can be a key part in digital products [12]. Therefore, we will link the benefits of cloud computing to the challenges of the product lifecycle stages in the next section.

The challenges of digital products across the product lifecycle can be linked to the benefits of cloud computing. During the introduction phase, uncertainty of market success can lead to a potentially unreliable forecast of required resources [17]. On the one hand, the benefit of high scalability of resources can lead to quick upscaling, in case demand is higher than expected. On the other hand, in case of an unsuccessful product, cloud computing allows for quick downscaling to free resources for more successful products, thus lowering overall financial exposure. In the same phase, the challenge with high up-front investment is a potential financial loss of any upfront investments made into product design and launch [19]. If cloud computing is employed to reduce the cost of market entry, the potential financial loss is also reduced.

Once the growth phase has been reached, a high risk of competition entering the market can be tackled by cloud computing's lower barrier to innovation. With its quick access to scaling resources and easy access to experiments, cloud computing can enable faster iterations of digital products, therefore allowing better reactions to customer needs and thus lowering the risk of losing customers to the competition.

This benefit can also help when facing slowing growth due to market saturation in the maturity phase: As market growth slows, companies will start competing for existing customers rather than trying to acquire new customers [19]. A high ability to innovate and iterate on a digital product will allow the company to differentiate its digital product from the competition. Additionally, cloud computing provides access to specialized technology, which is beneficial because as a market becomes saturated, digital products will require more unique and differentiating features [20]. These often come from new technological developments; therefore, cloud computing can be a supporting factor if it provides access to specialized technology that a company would not be able to provide internally. To tackle increased price pressure from competition, the ability to optimize for specific workloads becomes more important: As price pressure increases, a main goal is to reduce operating cost. Optimizing infrastructure for specific workloads will allow the infrastructure and transaction costs to decrease, therefore bringing the overall costs down.

Once a digital product enters the decline phase and faces shrinking market demand, so does the amount of infrastructure required. Scalable cloud computing resources allow the organization to re-purpose resources for other products and thus keep overall utilization of IT resources high.

4.3 A Conceptual Model for the Deployment of Cloud Computing

The relationship between product lifecycle stage, challenges, cloud computing benefits and deployment model described in previous paragraphs can be summarized in one unified model, linking the product lifecycle stage to potentially viable cloud deployment models using the challenges and benefits associated with them. The result of this consolidation is displayed in Table 2:

Table 2. Relationship of product lifecycle, cloud computing benefits and deployment model

Stage	Challenge	Cloud computing Benefit	Applicability of benefit			
			Private cloud	Community cloud	Public cloud	Hybrid cloud
Introduction	*Uncertainty of market success*	High scalability of resources	Medium	Medium	**High**	**High**
	High up-front investment	Lower cost of entry	Low	Medium	**High**	Medium
Growth	*High risk of competition entering the market*	Lower barrier for innovation	Medium	Medium	**High**	**High**
Maturity	*Slowing growth due to market saturation*	Lower barrier for innovation	Medium	Medium	**High**	**High**
		Access to specialized technology	Low	Medium	**High**	**High**
	Increased price pressure from competition	Ability to optimize for specific workload	**High**	Medium	Low	**High**
Decline	*Shrinking market demand*	High scalability of resources	Medium	Medium	**High**	**High**

For readability we have highlighted those deployment models that have a high applicability of the corresponding benefit. As is evident from the consolidated view, between one and two deployment models are dominant in each stage of the product lifecycle. These are:

1. Introduction: Public Cloud
2. Growth: Public Cloud or Hybrid Cloud (The product will likely remain on public cloud infrastructure during the growth stage, as no dominant benefit can be achieved by changing the deployment model now)
3. Maturity: Hybrid Cloud
4. Decline: Public Cloud or Hybrid Cloud

Given these dominant deployment models, we can deduct that an optimal path for introducing a digital product using cloud computing starts off with using public cloud resources to reduce the upfront investment. The product will likely remain on public cloud infrastructure during the growth stage, as no dominant benefit can be achieved by changing the deployment model now. At the maturity stage, price pressure and competition will increase the demand for more optimized computing resources, leading the organization to add private IT infrastructure, thus creating a hybrid cloud environment. In the decline phase, the organization will retain a hybrid cloud environment for a time, while scaling back resources and deploying them to other products. This can lead to either retaining the hybrid cloud infrastructure or, if cutting back on private IT resources first, can result in a move back to a purely public cloud environment. A graphical representation of this concept can be seen in Fig. 2.

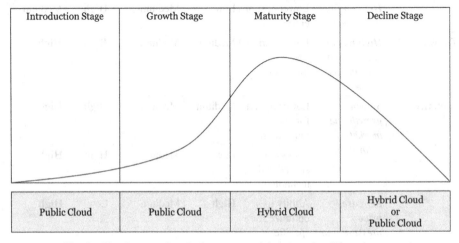

Introduction Stage	Growth Stage	Maturity Stage	Decline Stage
Public Cloud	Public Cloud	Hybrid Cloud	Hybrid Cloud or Public Cloud

Fig. 2. Cloud computing deployment models by product lifecycle stage

5   Limitations and Recommendations for Future Research

The conceptual model links cloud computing deployment models to stages of the product lifecycle for digital products. It provides a basis for determining the appropriate use of cloud computing when developing and maintaining a digital product. However, the model is based purely on the goal of achieving the highest benefit for developing a digital product. It does not consider other factors commonly associated with selecting a cloud computing deployment model. These factors include privacy, security or organizational factors like optimizing the utilization of existing IT infrastructure of the organization [23].

At first glance, our model suggests that private or community cloud deployment models are not relevant when developing new digital products. However, it must be acknowledged that any decision for or against cloud computing as a whole or one particular deployment model is always based on a multitude of factors, which are – as discussed previously - not all represented within our model. Reasons to choose private or community cloud models may include security concern, privacy requirements or legal/regulatory limitations [23].

We therefore encourage further development of the conceptual foundation, possibly resulting in the development of an advanced model, incorporating more of the dimensions mentioned above. Further research should be done to develop such a holistic model as a foundation for research into cloud deployment models and as a guidance for practitioners to select a cloud deployment model for their digital products. An additional research area is comprised of ways the critical shift from public to hybrid cloud environments, taking place between growth and maturity stage, can be facilitated by organizations as well as methodologies and tools that might support this shift. Lastly, empirical research into actual cloud computing usage in digital product development can help support the assumptions underlying our conceptual model and verify the validity of the suggested relation between product lifecycle stages and cloud computing deployment models.

6 Conclusion

Within this paper, we aim to introduce a conceptual model that links the benefits of cloud computing to the challenges of digital products across several stages of their product lifecycle. The focus, therefore, is on the different deployment models for cloud computing. We have covered the theoretical background of both cloud computing and the product lifecycle and have linked both concepts using by using an existing theory to explain a new phenomenon, as suggested by Yadav [14]. The contribution of this paper is twofold: We offer a theoretical concept for connecting the product lifecycle with cloud computing as a foundation for further research and discussion by other researchers, thus contributing to closing the gap identified by Senyo et al. [13]. Secondly, the paper provides a means for IT managers who are responsible for cloud computing in their organization for determining the appropriate use of cloud computing when developing or improving a digital product. Based on the model suggested, if no other factors like security, privacy, or regulatory concerns demand a private/community cloud environment, new digital products should be developed in a public cloud environment to reduce the risk and up-front investment associated with launching a product in a volatile market environment. Later, when cost reduction and optimization become more relevant in the maturity stage, a strategic shift towards a hybrid cloud environment is recommended to facilitate these optimizations using private cloud resources in selected areas of the product infrastructure.

References

1. Mell, P., Grance, T.: The NIST definition of cloud computing. https://nvlpubs.nist.gov/nis tpubs/Legacy/SP/nistspecialpublication800-145.pdf
2. IDG: 2018 cloud computing survey. https://www.idg.com/tools-for-marketers/2018-cloud-computing-survey/
3. Forni, A.A., van der Meulen, R.: Gartner says By 2020, a Corporate "No-Cloud" Policy Will Be as Rare as a "No-Internet" Policy Is Today. https://www.gartner.com/newsroom/id/335 4117
4. Bharadwaj, A., El Sawy, O.A., Pavlou, P.A., Venkatraman, N.: Digital business strategy: toward a next generation of insights. MIS Q. 37, 471–482 (2013)

5. Hanelt, A., Piccinini, E., Gregory, R.W., Hildebrandt, B., Kolbe, L.M.: Digital transformation of primarily physical industries - exploring the impact of digital trends on business models of automobile manufacturers. In: Wirtschaftsinformatik Proceedings 2015, p. 88 (2015). https://aisel.aisnet.org/wi2015/88
6. Vernon, R.: International investment and international trade in the product cycle. Q. J. Econ. **180**, 190–207 (1966)
7. Harik, R., Rivest, L., Bernard, A., Eynard, B., Bouras, A. (eds.): PLM 2016. IAICT, vol. 492. Springer, Cham (2017). https://doi.org/10.1007/978-3-319-54660-5
8. Liu, X., Jia, H., Guo, C.: Mobile application life cycle characterization via apple app store rank. Proc. Am. Soc. Inf. Sci. Technol. **51**, 1–4 (2014). https://doi.org/10.1002/meet.2014.14505101151
9. Nambisan, S. (ed.): Information Technology and Product Development. Springer, New York (2009). https://doi.org/10.1007/978-1-4419-1081-3
10. Dempsey, D., Kelliher, F.: Cloud computing: the emergence of the 5th utility. In: Dempsey, D. and Kelliher, F. (eds.) Industry Trends in Cloud Computing: Alternative Business-to-Business Revenue Models, pp. 29–43. Springer, Cham (2017). https://doi.org/10.1007/978-3-319-639 94-9_3
11. Liu, S., Chan, F.T.S., Yang, J., Niu, B.: Understanding the effect of cloud computing on organizational agility: an empirical examination. Int. J. Inf. Manage. **43**, 98–111 (2018). https://doi.org/10.1016/j.ijinfomgt.2018.07.010
12. Marston, S., Li, Z., Bandyopadhyay, S., Zhang, J., Ghalsasi, A.: Cloud computing—the business perspective. Decis. Support Syst. **51**, 176–189 (2011). https://doi.org/10.1016/j.dss.2010.12.006
13. Senyo, P.K., Addae, E., Boateng, R.: Cloud computing research: a review of research themes, frameworks, methods and future research directions. Int. J. Inf. Manage. **38**, 128–139 (2018). https://doi.org/10.1016/j.ijinfomgt.2017.07.007
14. Yadav, M.S.: The decline of conceptual articles and implications for knowledge development. J. Market. **74**, 1–19 (2010). https://doi.org/10.1509/jmkg.74.1.1
15. Mahmood, M.A., Arslan, F., Dandu, J., Udo, G.: Impact of cloud computing adoption on firm stock price – an empirical research. In: AMCIS 2014 Proceedings (2014). https://aisel.aisnet.org/amcis2014/GlobalIssues/GeneralPresentations/1
16. Litoiu, M., Woodside, M., Wong, J., Ng, J., Iszlai, G.: A business driven cloud optimization architecture. In: Proceedings of the 2010 ACM Symposium on Applied Computing - SAC 2010, Sierre, Switzerland, pp. 380–385. ACM Press (2010). https://doi.org/10.1145/1774088.1774170
17. Cox, W.E.: Product life cycles as marketing models. J. Bus. **40**, 375–384 (1967)
18. Rink, D.R., Swan, J.E.: Product life cycle research: a literature review. J. Bus. Res. **7**, 219–242 (1979). https://doi.org/10.1016/0148-2963(79)90030-4
19. Levitt, T.: Exploit the product life cycle (1965). https://hbr.org/1965/11/exploit-the-product-life-cycle
20. Al-Debi, M.M., El-Haddadeh, R., Avison, D.: Defining the business model in the new world of digital business. In: AMCIS 2008 Proceedings, p. 300 (2008). http://aisel.aisnet.org/amcis2008/300
21. App Annie Research: Annual number of mobile app downloads worldwide 2022|Statistic. https://www.statista.com/statistics/271644/worldwide-free-and-paid-mobile-app-store-downloads/
22. Strader, T.J., Lin, F.-R., Shaw, M.J.: Information infrastructure for electronic virtual organization management. Decis. Support Syst. **23**, 75–94 (1998). https://doi.org/10.1016/S0167-9236(98)00037-2
23. Chen, D., Zhao, H.: Data security and privacy protection issues in cloud computing. In: 2012 International Conference on Computer Science and Electronics Engineering, Hangzhou, Zhejiang, China, pp. 647–651. IEEE (2012). https://doi.org/10.1109/ICCSEE.2012.193

Antecedents of Different Social Network Structures on Open Source Projects Popularity

Shahab Bayati[✉] and Arvind Tripathi

The University of Auckland, Auckland, New Zealand
{s.bayati,a.tripathi}@auckland.ac.nz

Abstract. As Open source software (OSS) phenomenon become popular, it attracts millions of developers and plays a key role in success of small and large businesses. However, OSS ecosystem is very competitive and so only a few OSS projects, among millions hosted on social coding platforms such as GitHub, become successful. Since popular projects attract more developers, a key success ingredient, this research examines the antecedents of popularity of OSS projects hosted on a social coding platform. We have investigated the effect of the social structure of an OSS project on project popularity among community members. Data from GitHub is used to construct two different types of social networks for each project. The affiliation network represents the developers' inter-project relationships and following network reveals intra-project relationship. Applying the lenses of social network theory, we examine the effect of embeddedness and cohesion of the project's contributors on project popularity. Our results show that both affiliation and following networks are different in how they evolve and affect project popularity. Our findings can help OSS project leaders to understand developers' interactions and its effect on popularity of the project.

Keywords: Social coding · Social network theory · Open source software · Project popularity · Following network

1 Introduction

Open source community and artefacts are evolving and growing at a fast pace and new generations of OSS development are emerging (Fitzgerald 2006). This phenomenal growth of Open Source Software (OSS) over the last few decades has attracted both scholars and practitioners. With a large number of OSS projects, not all the projects survive or become successful and therefore scholars have investigated the drivers of success of OSS projects because project success is known as a good metric for project performance in general (Aksulu and Wade 2010). Researchers have studied different drivers of success of open source projects including license type, sponsorship, project maturity, level of activities, code base size, team size and team embeddedness (Grewal et al. 2006; Stewart et al. 2006; Subramaniam et al. 2009).

Along with these metrics, project popularity is also a success indicator especially for the projects that are not funded by well-known enterprises (Crowston et al. 2012;

© Springer Nature Switzerland AG 2020
K. R. Lang et al. (Eds.): WeB 2019, LNBIP 403, pp. 143–157, 2020.
https://doi.org/10.1007/978-3-030-67781-7_14

Stewart and Ammeter 2002). Popularity of a project is measured by end-users' interest, code changes, bug resolved and code quality, etc. (Cheruy et al. 2017; Ghapanchi et al. 2011). However, with the emergence of social coding platforms such as GitHub, the collaboration and communication patterns of OSS contributors are becoming transparent to OSS community (Dabbish et al. 2012) and provide new lenses for measuring popularity of OSS projects. Social coding platforms enhance the general OSS development repositories with social networking features (Moqri et al. 2018). The new features of social coding platforms enable researchers to evaluate the popularity of open source project with the new metrics such as the number of project watchers (Borges et al. 2016; Zerouali et al. 2019).

Applying the lenses of social network theory, scholars have investigated success of OSS projects through the structure and embeddedness of social ties among developers and projects (Grewal et al. 2006; Singh et al. 2011). Affiliation Network (Co-membership) is the most common type of network used in OSS literature (Peng 2019; Wu and Goh 2009). Various network related measures are used to investigate the effect of network ties on OSS project success (Grewal et al. 2006; Singh 2010; Wu and Goh 2009). However, we contend that working on the same project does not mean that developers have interest in each other's work, because they can work on different modules at different times. We believe that the unidirectional following event in social coding can elaborate more about developers' social interaction inside a project than a co-membership network. Developers follow each other to get the latest updates, find new opportunities and promote their social status (Blincoe et al. 2016). Therefore, we argue that a following network can realistically measure developers interests in each other work. In fact, researchers have used following networks in OSS to find collaboration patterns (Wu et al. 2014) and recommend developers to follow new peers (Schall 2014). Thus, this study aims to provide a more realistic perspective on developers' interactions by investigating the following network on social coding platforms.

We have compared the effects of both affiliation and following networks on the popularity of OSS projects among community members. Since OSS project is a dynamic ecosystem and developers' network structure changes over time (Wu and Goh 2009), we have also investigated the effect of changes in these networks on project popularity. Formally, our research questions are "RQ1: How developers' network dynamics affects the popularity of OSS projects across the community?" and "RQ2: How the affiliation network and the following network are different in their effect on project popularity?"

To empirically answer these research questions, we have collected a panel dataset of OSS projects and contributors for a period of 3 years. We have constructed two different networks of developers' interactions. First, following the literature, we have used the undirected network of co-membership. In this network, we have created a tie between two developers based on co-membership (both are member of a project). This network type mostly shows the external connection of developers and how they can expand the boundaries of a project. Second, based on the social networking feature of GitHub, we build a network for each project based on following event. In this case, there is a directed tie between developer "A" and "B" of a focal project, if developer "A" follows Developer "B". This following network demonstrates internal communication pattern of a project.

Internal and external OSS project popularity measures are reviewed in (Cheruy et al. 2017). The unit of analysis for this study is OSS project.

Drawing from social network theory, we have used the concept of centrality to evaluate the network structure (Grewal et al. 2006; Wu and Goh 2009). Centrality is measured at the network level and represents the importance and influence of a node in the entire network. In this study, we have considered the degree centrality, betweenness centrality, closeness centrality and eigenvector centrality. Degree centrality refers to direct ties with other developers, betweenness centrality evaluates the differences in brokerage power of developers in a network, closeness centrality reveals the variation in nodes distances from each other, and eigenvector centrality depicts how nodes are connected with important and core developers. In addition to these centrality measures, we have used density as another network metric, which represents the embeddedness of developers. These measurements are calculated for both network structures over time. We have estimated the effect of these network centrality measures on the popularity of OSS projects.

OSS success factors are generally divided into two main groups-internal and external success factors. Internal success factors refer to coding activities, project status, developers' skills, software quality and OSS norms and values (Grewal et al. 2006; Subramaniam et al. 2009), and external factors are mostly related to customer support and end-user satisfaction (Midha and Palvia 2012; Stewart et al. 2006). We have argued about importance of potential contributors' interest in a project. Potential contributors are not officially a project member but they are acting as a part of project community. We extend the literature of popularity by measuring potential contributors' interest in a project as a layer between project members and end-users.

This paper is structured as follows. The related literature is reviewed in the next section. The research model and the main research hypotheses are presented later. Data collection and sampling techniques are discussed afterwards. Data analysis and discussion are presented in the next section. Finally, the study is concluded and possible future studies are discussed.

2 Literature Review

Open source project performance can be evaluated through the success and popularity metrics of a project (Aksulu and Wade 2010). Different measurement factors are used with a various theoretical framework to investigate the OSS success antecedents (Cheruy et al. 2017; Ghapanchi et al. 2011; Jansen 2014; Mens et al. 2017). However, by the emergence of social coding platforms, new metrics are presented for popularity and success analysis across OSS community (Jarczyk et al. 2014; Mens et al. 2017; Zerouali et al. 2019). In literature, most of the success measures have relied on technical contribution inside the project or end-users' interests (Grewal et al. 2006; Midha and Palvia 2012; Subramaniam et al. 2009). However, we have argued project's ability to attract OSS community members who contribute to a project as outsiders, should also be considered as a measure of success. This study contributes by developing new measures of project popularity in social coding environment. We have measured popularity with different levels of granularity.

Information systems success (DeLone and McLean 1992) is one of the major theories used in OSS project success analytics. However, OSS literature argued about the unsuitability of this theory in OSS analysis (Crowston et al. 2006) because OSS projects are naturally different from the commercial artefacts as they are dominated with volunteers' contribution. Researchers have applied other theoretical lenses such as Coordination theory, theory of competency rallying (Crowston and Scozzi 2002; Ghapanchi and Aurum 2012), cue utilization theory (Midha and Palvia 2012), organizational theory (Temizkan and Kumar 2015), social capital (Hinds and Lee 2008), and social network theory (Grewal et al. 2006; Wu and Goh 2009) to study OSS project success. This study applies social network theory as a holistic view of an OSS project. This study investigates the effect of social network evolution on OSS project popularity over time.

Social network theory defines OSS community through different nodes and ties among developer and projects (Grewal et al. 2006; Singh et al. 2011) and social network analysis interprets a different aspect of communication patterns through a graph structure analysis. Thus, social network theory provides a whole picture of relationship embeddedness and mostly focuses on the community rather than individual attributes (Grewal et al. 2006; Singh et al. 2011). In OSS literature, social network theoretical lenses have been applied to see the effect of affiliation network (co-membership) on the project success in terms of code commitment (Singh 2010; Singh et al. 2011), task completion (Chou and He 2011), and success among end-users (Grewal et al. 2006). Social ties among developers and projects in affiliation network affect their project selection (Hahn et al. 2008) and therefore success rate in OSS community (Peng et al. 2013). A few studies have used other network structures, such as code dependencies to understand core-periphery patterns in the success of OSS projects (Amrit and Van Hillegersberg 2010). Watching events are used to investigate co-watching network effect on coding activities in GitHub (Peng 2019). The pull request submission and review process is used to construct a social network of OSS developers in GitHub which confirms the effect of pull-based network cohesion on pull request evaluation (El Mezouar et al. 2019). The affiliation network is a bidirectional network, however, the social coding environment has new features such as following which represent unidirectional ties. We contend that co-membership in a project cannot be interpreted as a direct social interaction among developers. Since the structure of these networks are different, we have investigated the effect of the following network on open source project success. We have compared following network with affiliation network antecedents on OSS project popularity. Prior work has compared effect of different networks in OSS literature (Peng 2019; Temizkan and Kumar 2015).

3 Research Model

The success and popularity of open source projects are studied with different perspectives (Aksulu and Wade 2010; Ghapanchi 2015). Popularity is generally tested through the page views, download count and subscribers (Cheruy et al. 2017; Stewart et al. 2006; Wu and Goh 2009). However, by introducing social networking features in OSS developments the popularity of a project among the community members can be investigated with other attributes such as watching (Dabbish et al. 2012; Jarczyk et al. 2014).

In GitHub, open source contributors can technically show their higher level of interest with forking or submitting a pull request in a project. We can consider these two factors as a proxy for technical popularity of a focal project among community members (Zerouali et al. 2019). In this study, we have measured these popularity metrics and we want to figure out the antecedents of popularity by applying social network theory. The research model of this study is presented in Fig. 1. The research model tests for the effect of changes across two different network types of OSS project's developers on the project popularity. Each network reflects a different aspect of a project. Affiliation network reveals inter-relationship of developers across the social coding platform. While the following network depicts the intra-relationship among developers inside a project.

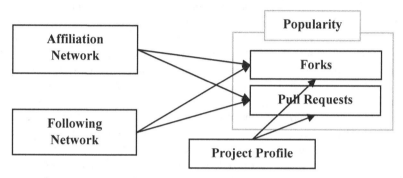

Fig. 1. Research model

Popularity construct is studied in two different levels of community members' interests as a proxy for project popularity across the community (Zerouali et al. 2019). Ranging from forking a project for possible modification and submitting pull-requests to contribute to a project as a non-member developer. These two measures of popularity are showing different levels of OSS community members' interests in a project. Compared to the literature, these measures illustrate the popularity among developers, while OSS literature measured popularity with the page-view, downloads (end-user level) and new members who joined (project level) (Subramaniam et al. 2009). In another words, this study defines a new middle tier to measure popularity across the community between project members and end-users.

The *Affiliation* network represents the contribution to other common projects outside of the projects. This construct defines how the project member in a focal project are tied to each other by considering their co-memberships in other projects. This type of network is most studied structure in OSS literature (Grewal et al. 2006; Singh 2010). Generally, affiliation network is used in literature to define the social interaction among OSS developers by constructing a network of developer tied to each other with their common projects. The affiliation network is a proxy for external cohesion of a project's developers (Singh et al. 2011). However, by applying social network theory we are testing its structural and embeddedness effect on OSS project popularity inside the community. In OSS, developers are generally volunteers and they contribute freely as much as they want in different project. Over time, more embedded affiliation network reveals that developers are joining other projects together and it may affect their level contribution in a focal project.

The *Following* network demonstrates interest of OSS developers inside a project. A tie is formed once a developer follow another one which has been used as a factor of internal cohesiveness of a project. Ignoring following network may lead to lose the direct social interaction among developers. Compared to affiliation network, the Following network is a unidirectional network. Over time, increasing, the number of following ties in a project shows more friendly and respectful environment. Following network is used to design developer/expert recommender systems in GitHub (Mo et al. 2015; Schall 2014). In GitHub mining literature these type of networks (affiliation and following) are studied together in (Mo et al. 2015).

Social coding platform interactions are transparent to everyone and developers inter-action inside a project are used to evaluate a project by community members (Dabbish et al. 2012). The direct effect of internal cohesiveness on the popularity of OSS projects inside a community is tested in this research model. To evaluate a project community members are seeking the project static and dynamic profile. Project profile is a construct used in this study to represent the project characteristics which can affect community members' evaluation of a focal project.

Although most of the applications of social network theory in OSS literature focused on affiliation network, we have argued other network structures such as the following network can show different relation to OSS popularity as nature is different. The fol-lowing network is a directed network which represents the internal communication of developers, while the affiliation network is undirected which shows the relationship of developers across other projects. Then these two networks are conceptually and naturally different and reveal different aspects of an OSS project. The internal and external social networks of developers feature request and bug resolution activities evolved differently over time in artefact development level (Temizkan and Kumar 2015). Internal cohesion of a project leads to higher performance and external cohesion diminish the performance. Internal cohesion improve the communication and knowledge sharing in a project, while external cohesion shows that project contributors are contributing in other projects and their focus may be lost from the focal project. Developers follow each other based on social reputation and more personal interest such as getting the latest updates (Blincoe et al. 2016). In addition, affiliation network can show the common interest in project level and following can show the interest in personal level. Based on the different nature of these networks we have argued that the affiliation network and the following network evolve in different manners over time.

H1. Affiliation networks social structure and cohesion evolution over time is different from the following network in a social coding platform.

Based on online and geographically distributed nature of OSS community, develop-ers lack the face to face communication. Most of the interactions are happening inside a community through the platform features. Internal cohesion of OSS developers facil-itates the information flow in a team (Temizkan and Kumar 2015). Developers follow each other to socialize, learn new concepts, getting latest updates, and finding new tech-nologies and trends (Blincoe et al. 2016). In an internally cohesive network information is easily flown through different paths which leads to the higher level of trustworthy and robustness in knowledge sharing (Schilling and Phelps 2007). Knowledge sharing and diffusion improves the accuracy of task coordination (Espinosa et al. 2007). Internally cohesive projects are more flexible to changes as the team is aware of norms and changes in a fast pace (Wu and Goh 2009). External resources are the source of innovation to a project, while developers working in other projects may learn new subjects and the focal project may get benefit of it. A cohesive affiliation network shows that most of developers have common working projects together which diminish novelty and innovation in a community (Temizkan and Kumar 2015). Redundant information is another issue raised in a high cohesive of external OSS network (Singh et al. 2011). Density represents how team members are tied to each other. A higher value for density shows the more developers are connected to the other ones. A denser affiliation network shows developers inside a project have more common projects which may lead to less effort for the focal project. The increasing trend of affiliation network density can be interpreted as the project contributors are attracted to other projects and lost their interest in the focal project (Daniel and Stewart 2016). On the other side, denser following network shows a more friendly community which may lead to higher socialization and commitment level among developers (Temizkan and Kumar 2015). Formally, we hypothesize that.

H2a. Considering the affiliation network, a denser project has a negative relationship on the popularity of OSS project over time.

H2b. Considering the following network, a denser project has a positive relationship on the popularity of OSS project over time.

High dependency on one or few developers in OSS projects leads to a high degree of centrality in its social network (Wu and Goh 2009). This situation may affect the survival of a project in the case of losing these core members (Wu and Goh 2009). Degree centrality of project developers shows how the direct power for information diffusion among developers is distributed across a project. A higher degree centrality means that some nodes are more powerful which indicates that power is not equally distributed among member. Low value for degree centrality is an indicator for power balance in a project The progressive evolution of nodes' power in a developers' network is not a good signal for outsiders and may be considered as a sign of a non-balanced community (Singh 2010). Same growth in betweenness centrality value in a network over time send a signal to the community that the specific nodes are coordinating and brokering information in a project. Closeness centrality also measures the distances among nodes, a higher value shows unequally distributed of information propagation speed in a network. Eigenvector centrality reveals the effect of connection with important nodes. Inequality of concentration of connection with important nodes signals to the community member that project members are evaluating other one's activities. Inequality in the affiliation network is a signal of the unbalanced contribution of developers. While inequality in the following network means of unbalanced knowledge flow and unequal power distribution in the focal project. These interoperation of centrality reveal that in both types of networks higher centrality can challenge the project popularity.

H3. Growth of network centrality is likely to be negatively correlated with open source project popularity among community members over time.

4 Data Collection and Pre-processing

We have collected a sample of longitudinal data of OSS projects hosted on GitHub. GitHub is the largest public OSS repository with more than 100 million projects and 40 million registered users. We have used GhTorrent 2016 MySQL data dump for this study which contains all public projects and developers' information since 2008. Since many projects on GitHub are personal, mirrored, and inactive projects (Kalliamvakou et al. 2014), we have applied several criteria for project retrieval. We have removed all forked and deleted projects and collected projects that were active (have coding activities in their lifespan), have at least five members (Nielek et al. 2016; Wu and Goh 2009), and . developed in one of the major programming languages (Bayati and Peiris 2018). To control for the maturity of the project, we have selected our sample from all the projects funded and created during the first quarter of 2013. We have quantified all the seasonal activities for a period of 3 years which is equivalent to 12 quarters. We have used seasonal data as the average time for minor releases of GitHub projects is close to 90 days (Yamashita et al. 2016). Our balanced panel dataset contains on 272 projects for 12 periods.

5 Measurements

In this study, we are focusing on project popularity inside the community. Although many measures such as commit level, bug fixed, subscribers and download rates are widely used in OSS literature, we have used two different measures-number of forks and pull requests, to capture popularity among developers. Number of forks measures number of times the project's code has been copied for further changes and modifications. The changes on a forked copy may be aligned with the main project or in a new direction (Rastogi and Nagappan 2016). This feature can show the popularity of project among community members who wish to make technical contribution. A Pull Request can reveal the OSS contributors interests to pull back their contribution to the main project. Pull requests reflects the community interests to promote and contribute to the focal project. We have argued that these two measures allows us to capture project popularity among developers' community. The list of variables used in this study are described in Table 1.

We have presented the outcome of correlation analysis in Table 2. In this study, we are using a longitudinal dataset and general correlation approaches may be affected with the auto-correlated values over time. As it has been suggested in (Gelman and Hill 2007) we have presented the between-subject correlation matrix.

Based on the correlation matrix analysis, the overall number of pull requests among projects is correlated with forks as both are representing popularity of a project. Across dependent variables, betweenness centralities for both networks are moderately correlated with closeness centrality. To check for the multi-collinearity issue we have used VIF measure across both popularity measures and all values were <2.0 and it will not affect panel data analysis. This pairwise correlation analysis can give us a good understanding of our dataset.

Table 1. Variables description

Variable	Description
Forks	Number of times the focal project is forked at time t
PullReqs	Number of pull requests submitted to the focal project at time t
Members	Number of project core members at time t
LT	Project License Type
DensFol	Density of Following network in the focal project at time t
DensAff	Density of affiliation network in the focal project at time t
ColCenFol	Closeness Centrality of Following network in the focal project at time t
ColCenAff	Closeness Centrality of affiliation network in the focal project at time t
BetCenFol	Betweenness Centrality of Following network in the focal project at time t
BetCenAff	Betweenness Centrality of affiliation network in the focal project at time t
EgnCenFol	Eigenvector Centrality of Following network in the focal project at time t
EgnCenAff	Eigenvector Centrality of affiliation network in the focal project at time t
DegCenFol	Degree Centrality of Following network in the focal project at time t
DegCenAff	Degree Centrality of affiliation network in the focal project at time t

Table 2. Correlation matrix among projects (Between-Subject)

	1	2	3	4	5	6	7	8	9	10	11	12	13
1. Forks	1												
2. PullReqs	0.64	1											
3. Members	0.01	0.03	1										
4. DensFol	0.13	−0.01	−0.26	1									
5. DensAff	−0.16	−0.07	0.19	0.01	1								
6. CloCenFol	−0.01	−0.01	−0.20	0.62	0.09	1							
7. CloCenAff	0.12	0.05	−0.18	0.13	−0.26	−0.01	1						
8. BetCenFol	−0.01	−0.01	−0.07	0.69	0.09	0.58	0.01	1					
9. BetCenAff	0.12	−0.01	−0.10	0.14	−0.20	0.02	0.70	0.04	1				
10. EgnCenFol	−0.02	0.04	0.31	0.32	0.25	0.35	−0.09	0.34	−0.03	1			
11. EgnCenAff	0.26	0.13	−0.05	0.04	−0.54	−0.07	0.58	−0.07	0.45	−0.08	1		
12. DegCenFol	−0.02	−0.00	−0.00	0.54	0.18	0.81	−0.06	0.48	−0.00	0.56	−0.12	1	
13. DegCenAff	0.24	0.16	−0.12	0.08	−0.45	−0.06	0.84	−0.04	0.66	−0.05	0.85	−0.09	1

6   Panel Data Analysis and Discussion

This section depicts the results of panel data analysis. We have modelled our analysis based on Random-Effect econometrics model as the License Type (LT) variable is time-invariant (Subramaniam et al. 2009). Based on the normality check analysis, some of the variables are skewed including all the dependent variables and the network related metrics. Logarithmic transformation is suggested in (Gelman and Hill 2007) for data normalization in these cases. We have presented the coefficient estimates of two models analysis based on the outcome of R PLM package (Croissant and Millo 2008) in Table 3. Furthermore, we have discussed the major findings and implications of this analysis for OSS researchers and practitioners.

Table 3. Panel data analysis results

	Forks$_t$		Pull requests$_t$	
	Coef. (Std Err)	P-Value Sig	Coef. (Std Err)	P-Value Sig
Intercept	−0.066 (0.206)	0.746	0.527 (0.298)	0.076 *
Ln(DensAff$_t$)	−0.204 (0.021)	0.000 ***	−0.273 (0.030)	0.000 ***
Ln(DensFol$_t$)	0.166 (0.063)	0.008 ***	0.052 (0.088)	0.549
Ln(DegCenAff$_t$)	−0.040 (0.070)	0.559	−0.294 (0.097)	0.002 ***
Ln(DegCenFol$_t$)	−0.193 (0.049)	0.000 ***	−0.193 (0.069)	0.005 ***
Ln(BetCenAff$_t$)	−0.074 (0.028)	0.008 ***	−0.118 (0.039)	0.002 ***
Ln(BetCenFol$_t$)	0.022 (0.034)	0.524	0.022 (0.048)	0.641
Ln(CloCenAff$_t$)	0.038 (0.067)	0.575	0.371 (0.094)	0.000 ***
Ln(CloCenFol$_t$)	−0.053 (0.049)	0.279	−0.086 (0.069)	0.211
Ln(EgnCenAff$_t$)	0.028 (0.029)	0.338	0.029 (0.040)	0.469
Ln(EgnCenFol$_t$)	−0.027 (0.015)	0.078 *	−0.007 (0.021)	0.738
LT$_{Permissive}$	0.669 (0.214)	0.001 ***	1.537 (0.323)	0.000 ***
LT$_{StronglyProtective}$	0.351 (0.335)	0.294	0.587 (0.507)	0.247
LT$_{WeaklyProtective}$	0.052 (0.437)	0.903	−0.050 (0.661)	0.938
LT$_{Others}$	0.801 (0.216)	0.000 ***	1.310 (0.327)	0.000 ***
Ln(Members$_t$)	0.943 (0.069)	0.000 ***	0.954 (0.099)	0.000** *
Adj. R−Squared	0.116		0.091	
Idiosyncratic Share	0.280		0.539	
Individual Share	1.388		3.218	
AIC	5115.8		7172.5	

Signif. codes: 0.01 '***' 0.05 '**' 0.1 '*'

6.1 Social Network Related Findings

Results in Table 3 reveal that there is a significant and negative correlation between the density of affiliation network and popularity measures. Also, regarding the following network density and popularity the model shows a positive and significant correlation with forks and non-significant with pull requests.

According to centrality measures, degree centrality measures in the following network has a negative and the significant effect on popularity measures over time while the only significant correlation in the affiliation network is found in pull request which is also negative. Eigenvector centrality does not show the significant outcome. Betweenness in the affiliation network plays a negative significant role in relation to popularity metrics. This fact reveals that in the affiliation network, moving forward to more equal network in terms of brokerage power makes the project more popular. Closeness in affiliation network, has a positive and significant effect on pull requests which reverts our arguments. Inequality in distance among developers in affiliation positively affects project popularity. Equal and dens network can be a signal for an outsider that a team is already shaped as they have many common projects and there is a low potentiality for other contributions. Regarding H1, to compare two networks through their associated measures, we have followed (Temizkan and Kumar 2015). Using Wilcoxon test we found the significant differences among all metrics in two networks which supports our hypothesis.

Generally, the analysis of network features reveals that the affiliation network and the following networks regulate different effects on OSS project popularity. In the case such as density, their effect is completely reverse which supports H1 and H2. In other cases, where we can show a significant effect on one network the other one does not show statistically significant results. The analysis approves that density of internal social network (following) is positively correlated with project popularity (H2b). In contrast, regarding the H2a there is a negative correlation between popularity factors and external network (affiliation network). These confirms our arguments about social network density. According to H3, the current outcome partially supports the hypothesis, however, in few cases such as closeness it approves a completely reverse idea.

6.2 Project Characteristics Findings

The longitudinal model depicts the significant relationship of joining new members to the project team on the popularity of a project over time. Accepting new members significantly send signals to the community for the openness of the project for new contributions. Regarding license choice, projects with permissive licences are getting more popular over time. Same results are shown regarding projects with non-general licenses. The effect is higher for permissive type on pull requests. However, fork-ability of no-general licenses is higher than permissive ones over time (Table 4).

These findings can help OSS project managers and owners to keep track of developers' interaction inside a project as all of them are considered as a signal to the community. Intellectual property rights such as license choice required to be chosen very well in terms of project goals. Also, the democratic organizational model with power equality over time direct project to more popularity inside the community. On the other side, it provides an overview for community members and developers to understand the nature of interaction better. Project members can realize how their internal and external socio-technical interactions can affect the project reputation.

Table 4. Summary of the analysis of hypotheses

Hypothesis		Results
H1	Affiliation networks social structure and cohesion evolution over time is different from the following network on a social coding platform.	Supported
H2a	Considering the affiliation network, a denser project has a negative relationship on the popularity of OSS project over time.	Supported
H2b	Considering the following network, a denser project has a positive relationship on the popularity of OSS project over time.	Supported
H3	Growth of network centrality negatively correlated with open source project popularity among community members over time.	Partially Supported

6.3 Limitations

Along with the original practical outcomes of this study, we are aware of a few limitations in this paper. First, our study relies on GhTorrent as the main source of data. There may be issues in the way GhTorrent collect and demonstrate data. However, GhTorrent is widely used in GitHub mining studies and most of the studies rely on it, Second, we have tested our model and hypotheses based on the random subset of GitHub projects. Although the sample is big enough for a longitudinal data analysis, however, the same analysis can be applied to different datasets to provide more robust insights. Third, although GitHub is the largest social coding community for OSS development, other communities such as BitBucket and GitLab can be used for platform effect analysis. Fourth, this study only covers the activities inside the GitHub community, however, some projects have other communication channels such as slack that we have missed them. Fifth, regarding pull requests, following the literature, we have only counted the number of pull requests (Sarker et al. 2019) but in reality pull requests are different in size and effect. A pull request may contains one or more code commits while each commit can vary in its size and quality. In addition, pull requests' final status is also important (merged/rejected).

7 Conclusion and Future Works

In this study, we have compared two different social networks structures of OSS developers in OSS projects. We have tested a series of hypotheses regarding OSS project popularity. Panel data analysis is applied on 272 projects hosted on GitHub. Affiliation and following network are structured for each project for a period of three years. We have calculated different network related metrics on each project over time. The result of the longitudinal model supports most of our arguments about differences in networks' behaviours and their effect on OSS project popularity. In addition, we have tested the effect of OSS licence and the team size on project popularity over time. Permissive licences increase the chance of popularity.

For future research trend, we are planning to run the same model on other project success metrics. In addition, a project popularity monitoring artefact can be designed to evaluate and predict the network structure of developers based on link prediction methods. Furthermore, other possible networks such as co-committing, co-watching and co-discussed can be shaped to study the project evolved over time.

References

Aksulu, A., Wade, M.: A comprehensive review and synthesis of open source research. J. Assoc. Inf. Syst. **11**(11), 576 (2010)

Amrit, C., Van Hillegersberg, J.: Exploring the impact of socio-technical core-periphery structures in open source software development. J. Inf. Technol. **25**(2), 216–229 (2010)

Bayati, S., Peiris, K.: "Road to Success: How Newcomers Gain Reputation in Open Source Community," PACIS 2018. AIS, Japan (2018)

Blincoe, K., Sheoran, J., Goggins, S., Petakovic, E., Damian, D.: Understanding the popular users: following, affiliation influence and leadership on github. Inf. Softw. Technol. **70**, 30–39 (2016)

Borges, H., Hora, A., Valente, M.T.: Predicting the popularity of github repositories. In: Proceedings of the 12th International Conference on Predictive Models and Data Analytics in Software Engineering, ACM, p. 9 (2016)

Cheruy, C., Robert, F., Belbaly, N.: Oss popularity: understanding the relationship between user-developer interaction, market potential and development stage. Systèmes d'information Management **22**(3), 47–74 (2017)

Chou, S.W., He, M.Y.: The factors that affect the performance of open source software development–the perspective of social capital and expertise integration. Inf. Syst. J. **21**(2), 195–219 (2011)

Croissant, Y., Millo, G.: Panel data econometrics in R: the Plm package. J. Stat. Software **27**(2), 1–43 (2008)

Crowston, K., Howison, J., Annabi, H.: Information systems success in free and open source software development: theory and measures. Software Process: Improvement and Practice **11**(2), 123–148 (2006)

Crowston, K., Scozzi, B.: Open source software projects as virtual organisations: competency rallying for software development. IEE Proceedings-Software **149**(1), 3–17 (2002)

Crowston, K., Wei, K., Howison, J., Wiggins, A.: Free/Libre open-source software development: what we know and what we do not know, ACM Computing Surveys (CSUR), **44**(2), 7 (2012)

Dabbish, L., Stuart, C., Tsay, J., Herbsleb, J.: Social coding in github: transparency and collaboration in an open software repository. In: Proceedings of the ACM 2012 Conference on Computer Supported Cooperative Work: ACM, pp. 1277–1286 (2012)

Daniel, S., Stewart, K.: Open source project success: resource access, flow, and integration. The J. Strategic Inf. Syst. **25**(3), 159–176 (2016)

DeLone, W.H., McLean, E.R.: Information systems success: the quest for the dependent variable. Inf. Syst. Res. **3**(1), 60–95 (1992)

El Mezouar, M., Zhang, F., Zou, Y.: An Empirical Study on the Teams Structures in Social Coding Using Github Projects, Empirical Software Engineering, pp. 1–34 (2019)

Espinosa, J.A., Slaughter, S.A., Kraut, R.E., Herbsleb, J.D.: Team knowledge and coordination in geographically distributed software development. J. Manage. Inf. Syst. **24**(1), 135–169 (2007)

Fitzgerald, B.: The transformation of open source software, MIS Quarterly, 587–598 (2006)

Gelman, A., Hill, J.: Data Analysis Using Regression and Multilevel/Hierarchical Models. Cambridge University Press, Cambridge (2007)

Ghapanchi, A.H.: Investigating the interrelationships among success measures of open source software projects. J. Organ. Comput. Electron. Commerce **25**(1), 28–46 (2015)

Ghapanchi, A.H., Aurum, A.: Competency rallying in electronic markets: implications for open source project success. Electron. Markets **22**(2), 117–127 (2012)

Ghapanchi, A.H., Aurum, A., Low, G.: Taxonomy for Measuring the Success of Open Source Software Projects, First Monday, 16(8) (2011)

Grewal, R., Lilien, G.L., Mallapragada, G.: Location, Location, Location: how network embeddedness affects project success in open source systems. Manage. Sci. **52**(7), 1043–1056 (2006)

Hahn, J., Moon, J.Y., Zhang, C.: Emergence of new project teams from open source software developer networks: impact of prior collaboration ties. Inf. Syst. Res. **19**(3), 369–391 (2008)

Hinds, D., Lee, R.M.: Social network structure as a critical success condition for virtual communities. In: Proceedings of the 41st Annual Hawaii International Conference on System Sciences (HICSS 2008), IEEE, p. 323 (2008)

Jansen, S.: Measuring the health of open source software ecosystems: beyond the scope of project health. Inf. Software Technol. **56**(11), 1508–1519 (2014)

Jarczyk, O., Gruszka, B., Jaroszewicz, S., Bukowski, L., Wierzbicki, A.: Github projects. quality analysis of open-source software. In: International Conference on Social Informatics, Springer, pp. 80–94 (2014)

Kalliamvakou, E., Gousios, G., Blincoe, K., Singer, L., German, D.M., Damian, D.: The promises and perils of mining github. In: Proceedings of the 11th Working Conference on Mining Software Repositories, ACM, pp. 92–101 (2014)

Mens, T., Adams, B., Marsan, J.: Towards an Interdisciplinary, Socio-Technical Analysis of Software Ecosystem Health, arXiv preprint arXiv:1711.04532 (2017)

Midha, V., Palvia, P.: Factors affecting the success of open source software. J. Syst. Software **85**(4), 895–905 (2012)

Mo, W., Shen, B., He, Y., Zhong, H.: Geminer: mining social and programming behaviors to identify experts in github. In: Proceedings of the 7th Asia-Pacific Symposium on Internetware, ACM, pp. 93–101 (2015)

Moqri, M., Mei, X., Qiu, L., Bandyopadhyay, S.: Effect of "Following" on contributions to open source communities. J. Manage. Inf. Syst. **35**(4), 1188–1217 (2018)

Nielek, R., Jarczyk, O., Pawlak, K., Bukowski, L., Bartusiak, R., Wierzbicki, A.: Choose a job you love: predicting choices of github developers. In: 2016 IEEE/WIC/ACM International Conference on Web Intelligence (WI): IEEE, pp. 200–207 (2016)

Peng, G.: Co-membership, networks ties, and knowledge flow: an empirical investigation controlling for alternative mechanisms. Decision Support Syst. **118**, 83–90 (2019)

Peng, G., Wan, Y., Woodlock, P.: Network ties and the success of open source software development. The J. Strategic Inf. Syst. **22**(4), 269–281 (2013)

Rastogi, A., Nagappan, N.: Forking and the sustainability of the developer community participation–an empirical investigation on outcomes and reasons. In: 2016 IEEE 23rd International Conference on Software Analysis, Evolution, and Reengineering (SANER), pp. 102–111. IEEE (2016)

Sarker, F., Vasilescu, B., Blincoe, K., Filkov, V.: Socio-technical work-rate increase associates with changes in work patterns in online projects. In: ICSE 2019, Canada (2019)

Schall, D.: Who to follow recommendation in large-scale online development communities. Inf. Softw. Technol. **56**(12), 1543–1555 (2014)

Schilling, M.A., Phelps, C.C.: Interfirm collaboration networks: the impact of large-scale network structure on firm innovation. Manage. Sci. **53**(7), 1113–1126 (2007)

Singh, P.V.: The small-world effect: the influence of macro-level properties of developer collaboration networks on open-source project success. ACM Trans. Software Eng. Methodol. (TOSEM), **20**(2), 6 (2010)

Singh, P.V., Tan, Y., Mookerjee, V.: Network effects: the influence of structural capital on open source project success, MIS Quarterly, **21**, 813–829 (2011)

Stewart, K., Ammeter, T.: An exploratory study of factors influencing the level of vitality and popularity of open source projects, In: ICIS 2002 Proceedings, p. 88 (2002)

Stewart, K.J., Ammeter, A.P., Maruping, L.M.: Impacts of license choice and organizational sponsorship on user interest and development activity in open source software projects. Inf. Syst. Res. **17**(2), 126–144 (2006)

Subramaniam, C., Sen, R., Nelson, M.L.: Determinants of open source software project success: a longitudinal study. Decision Support Syst. **46**(2), 576–585 (2009)

Temizkan, O., Kumar, R.L.: Exploitation and exploration networks in open source software development: an artifact-level analysis. J. Manage. Inf. Syst. **32**(1), 116–150 (2015)

Wu, J., Goh, K.Y.: Evaluating longitudinal success of open source software projects: a social network perspective, In: Hicss, IEEE, pp. 1–10 (2009)

Wu, Y., Kropczynski, J., Shih, P. C., Carroll, J.M.: Exploring the ecosystem of software developers on github and other platforms. In: Proceedings of the Companion Publication of the 17th ACM Conference on Computer Supported Cooperative Work & Social Computing, pp. 265–268. ACM (2014)

Yamashita, K., Kamei, Y., McIntosh, S., Hassan, A.E., Ubayashi, N.: Magnet or sticky? measuring project characteristics from the perspective of developer attraction and retention. J. Inf. Process. **24**(2), 339–348 (2016)

Zerouali, A., Mens, T., Robles, G., Gonzalez-Barahona, J.M.: On the diversity of software package popularity metrics: an empirical study of Npm. In: 2019 IEEE 26th International Conference on Software Analysis, Evolution and Reengineering (SANER), pp. 589–593. IEEE (2019)

Language Alternation in Online Communication with Misinformation

Lina Zhou[1], Jaewan Lim[1], Hamad Alsaleh[1], Jieyu Wang[2], and Dongsong Zhang[1(✉)]

[1] The University of North Carolina at Charlotte, Charlotte, NC 28223, USA
{lzhou8,jlim13,halsale2,dzhang15}@uncc.edu
[2] St. Cloud State University, St. Cloud, MN 56301, USA
jwang2@stcloudstate.edu

Abstract. Misinformation has attracted widespread research attention owing to the importance and challenges of addressing the problem. Despite the progress in developing computational methods for misinformation detection and identifying psycholinguistic factors and network properties of misinformation, there is a lack of understanding of language alternation in online misinformation communication. This research aims to address the literature gap by answering several key questions regarding language alternation in communicating misinformation online. Based on the analysis of structured and un-structured survey responses, the results of this study show that it is fairly common for speakers of English as a second language to alternate languages in communicating misinformation online. In addition, it identifies the motives and impacts of such behavior. The findings of this study point to a new avenue for addressing the challenges of detecting online misinformation.

Keywords: Misinformation · Online communication · Language alternation · Speaker of English as a second language

1 Introduction

The notion of misinformation is not new, which has attracted increasing amounts of research and media attention over the past decade. This is partly because the ubiquity of digital and networking technologies has fueled the growth and diffusion of misinformation. In particular, misinformation can be easily created, duplicated, shared, and distributed over a variety of online platforms such as social media websites and online discussion forums. The availability of considerable online information can precipitate rumors and falsified opinions [1]. In addition, these online platforms have increasingly become the primary source of information for online users. If people repeatedly receive unsubstantiated information, they are more likely to believe the conspiracy and be vulnerable to misinformation due to social reinforcement [2]. Furthermore, online users are subject to continuous exposure to misinformation such as fake news and false claims and become victimized or even unknowingly contribute to the spread of misinformation. Based on a large-scale evaluation of the echo chamber effect on political fake news

K. R. Lang et al. (Eds.): WeB 2019, LNBIP 403, pp. 158–168, 2020.
https://doi.org/10.1007/978-3-030-67781-7_15

consumption [3], approximately 1 in 4 Americans visited a fake news website between October 7 and November 14, 2016. Another study of fake news on Twitter during the 2016 U.S. presidential election found that fake news accounted for nearly 6% of all news consumption [4]. In practice, intentionally fabricated information can cause economic loss, politically biased belief, polarized social opinion, and defaming someone's reputation [5–7]. Therefore, it is crucial to understand online misinformation behavior.

There is an increasing body of literature on misinformation. For instance, one stream of research is focused on misinformation detection [8], and another on understanding the psychological and social factors with regard to misinformation assessment (e.g., [9]). The findings of these studies demonstrate that the identification of misinformation is a complex problem, which requires multidisciplinary efforts and should be approached after understanding the influence of a variety of contextual factors. However, previous studies have overlooked the influence of language background of information senders and the behavior of alternating between one's native and a foreign language in communicating misinformation. To address this significant limitation of the literature, this study investigates language alternation behavior in communicating misinformation online.

Language alternation can be driven by various motivations. For instance, sociolinguists have studied language alternation because of its benefit to and influence on promoting solidarity among individuals of different groups, persuasion of ideas, and credibility and reliability among the audience. Language alternation has been investigated extensively in education environments [10, 11], which can help teachers in foreign language education better communicate and motivate their students to participate. Language alternation has also been investigated in topic discussion, as people in different communities are used to discuss taboo words in a language different than their native language because of the ease and comfort in delivering their emotions and drawing attention [12]. Thus, it is reasonable to assume that language alternation is exploited in communicating misinformation online. However, this issue has yet to be examined in the literature.

This study aims to explore the following research questions in this nascent area: 1) Do speakers of English as a second language alternate languages when they communicate misinformation online? 2) Which language(s) do speakers of English as a second language prefer to use when they communicate misinformation online? 3) What are the motivations for alternating languages in communicating misinformation online? and 4) Who are more likely to be targeted by online misinformation that involves language alternation, speakers of English as a second language or native English speakers?

The rest of the paper is organized as follows. We introduce the background and review related work on misinformation and language alternation in the next section. Subsequently, we present a user study designed to collect data to answer the research questions. Finally, we report the results of data analyses, and discuss the findings and research contributions and implications.

2   Background and Related Work

2.1   Online Misinformation and Its Detection

Misinformation is defined broadly as disseminating deliberate deception, low-quality information, or hyper partisan news [5]. The disentanglement of features of information and misinformation can be conducted along five dimensions, including authority, accuracy, objectivity, currency, and coverage [13]. Compared with information, misinformation is not obvious about who creates its content and does not involve data to support accuracy. Additionally, misinformation typically contains subjective opinion, incomplete metadata such as information source, time, and place).

Aiming to detect online misinformation and its dissemination, existing studies can be categorized into four focused areas: 1) developing computational methods for detecting misinformation [8, 14], 2) identifying network properties of misinformation diffusion [15, 16], 3) crowdsourcing to flag suspicious social media postings [17, 18], and 4) psychology-based credibility assessment [9]. We illustrate each type of research using a representative study as the following:

- Computational methods for misinformation detection: Zhang et al. (2019) proposed a natural language processing framework to detect fake news under the assumption that fake news credibility can be measured by analyzing the credibility of the sentences within news [8]. A single sentence in a news article (regardless of fake or not) depicts an event linguistically with the subject, object, and predicate parts. The title of a news article is a combination of the subject and object. Accordingly, fake news can be detected through either fake topic detection that compares them against authentic news corpus or fake predicate detection that uses verb lists obtained from legitimate articles. Their model was able to achieve an accuracy of 92.49%.
- Network properties of misinformation diffusion: Misinformation diffusion patterns in the Twitter network is conducive to understanding internal and external activities when intelligent Twitter bots disseminate misinformation through interacting with human users [16]. The pattern starts with a bot's searching and crawling target information, followed by spreading misinformation internally to his Twitter followers. In a subsequent step, misinformation is dispersed externally through their followers' retweet or mentioning the tweets created by the bot to their own followers. By this way, misinformation can be distributed and augmented by illegitimate bots, consequently affecting external users and internal followers.
- Crowdsourcing to flag suspicious social media postings: DETECTIVE [18] utilized the power of crowdsourcing and Bayesian inference to curb the diffusion of misinformation. Its underpinning is one of Facebook's recent features where each user can flag a post as fake if he/she finds malicious news or claims. Using this self-reporting tool, the researchers developed an algorithm to label misinformative postings and sent them to experts for manual verification. By doing so, the researchers demonstrate the effectiveness of the algorithm in detecting fake news among Facebook users.
- Psychology-based credibility assessment: Kumar and Geethakumari (2014) characterized credible information with four features [9]. The first feature is the consistency of information, which can be determined based on whether information is compatible

and consistent with what we believe. The second is information coherence, which is concerned with whether a message is internally coherent or not. The third feature is the credibility of the information source, and the last feature is the acceptability and believability from others.

While previous misinformation studies have demonstrated significant initial success, they have overlooked the influence of language choice, particularly language alternation, on communicating misinformation online.

2.2 Language Alternation in Online Communication

Language alternation occurs when a speaker alternates between two or more languages in the same discourse. Social linguistic theories have been developed to explain language alternation behavior (e.g., [19]). There is a stream of research on the reasons and motivations for language alternation, such as quoting someone and solidarity and gratitude [20]. However, extant studies have primarily focused on language alternation in face-to-face communication. The research on the phenomenon in online communication remains scarce.

Among the very few studies that have touched upon language alternation in online communication, one study examined the reasons for language alternation in asynchronous communication by analyzing Facebook status updates wall [21]. The study revealed several factors that motivated online users to alternate language, including mentioning quotes in another language, addressing specification such as thanking someone in their own language, qualification for amplifying their posts by alternating to another language, checking and confirming opinions from others, and expressing emotions. Based on another study of language alternation in blogs [22], Italian expats in Netherlands alternate between languages for exhibitionism. Italians tend to show they are multilingual in order to demonstrate that they can adapt well to a new environment. Anglicism (easier to remember words in another language) is another factor that has been mentioned in literature [21]. Despite their overlap, some of the above reasons are unique to online communication that are not yet identified from face-to-face communication.

Gender factor has been examined in relation to language alternation [23]. It was found that approximately 80% of the participants alternated languages while using online social networking sites. Compared with females whose language alternation accounted for 60% of the population, the percentage of their male counterparts was 28%. Those findings were opposite to the previous finding from face-to-face communication [24]. The latter study shows that males tend to alternate languages more often than females in a mixed conversational context. Taking the above findings as a whole, they suggest that language alternation behaviors in online communication may differ from those in face-to-face communication. In other words, it is important and necessary to investigate language alternation in online communication separately.

Despite the very limited literature on language alternation in online communication, our literature review suggests that language alternation in online communication might be driven by different motivations from those in face-to-face communication. In addition, studies directly addressing language alternation for online misinformation communication remain scarce. This study will fill the void by investigating language alternation in online misinformation communication.

3 Method

We conducted a lab experiment for collecting data to answer the proposed research questions. At the beginning of the experiment, the participants were asked to respond to a survey questionnaire about their demographic information, language background, experience with online communication, and experience with language alternation as an information sender (who knowingly create and transmit misinformation) and as a receiver (who is the target of misinformation). Then, they were asked to create misinformation under different research settings. The focus of this research is on the structured and unstructured survey responses. For data analysis, we used a combination of quantitative and qualitative methods.

3.1 Participants

We recruited participants from a university located in the southern region of the U.S. University is an appropriate environment for recruiting participants for this study because a significant percentage of university students and employees are international and are adopters of online communication tools. The study was conducted in a research lab. There was a total of 31 participants who completed the entire study. After filtering out the participants who were not found to answer the survey questions seriously and/or used online communication tools frequently (less than once per week), and those who spoke another language other than English at home but have been staying in the U.S. for less than 3 months, we are left with 29 participants. Among them, 14 self-claimed English as their second language, and the rest were native English speakers who spoke at least one second language.

The participants who had English as a second language had diverse backgrounds in terms of their native language, including Arabic, Chinese, French, Hindi, Korean, and Spanish. The participants consisted of 6 full-time university employees and the rest were graduate or undergraduate students. Female accounted for 65.5%. The participants ranged from 18 to 39 years of age (mean $= 24.6$, std $= 6.15$).

Among the participants, 16 (55%) have used an online communication tool such as email or Google Hangouts on a personal computer several times a day, and all participants except for two used a tool on a daily basis over the past year. As far as the time spent on online communication using a personal computer is concerned, six spent 4 h or more (15.4%), ten (34.5%) 2 to 4 h, and eight (27.6%) 1 to 2 h, respectively.

3.2 Research Instrument and Variables

Given the lack of related prior studies, we created the main questionnaire items from scratch. Those questions covered two perspectives of online communication. From a perspective of misinformation sender, the participants were asked about their intention to alternate languages in communicating misinformation online to strangers and to family and friends, and their intention of using English during the communication. The questionnaire items were measured using a 7-point Likert scale, with 1 representing 'strongly disagree', 4 representing 'neural', and 7 representing 'strongly agree'. We included a couple of trap questions to screen for noisy responses.

To gain deep insights into the motives of alternating languages in communicating misinformation online, we further asked several open-ended questions about their own motives for alternating languages and their perceived motives of others for alternating languages when sending misinformation online. The questionnaire was administrated online.

There were three dependent variables, including language alternation in sending misinformation (LA), English use in sending misinformation (EG), and language alternation in received misinformation (RE_LA).

Speaker type (i.e., speaker of English as a second language vs. native English speaker) was the independent variable. Since a participant could be both the receiver and the sender of misinformation, so speaker type was used to categorize senders and receivers depending on the context of questions.

4 Analyses and Results

We used both quantitative and qualitative methods to analyze the data. We will introduce the results of quantitative and qualitative analyses separately.

4.1 Quantitative Results

To investigate whether the participants who speak English as a second language alternate languages when they communicate misinformation online, we performed a one-sampled t-test of LA_ST (mean $=$ 5.14, std $=$ 1.875) against the mid-point of 4. The results show that speakers of English as a second language alternate languages when they communicate misinformation online ($p < .05$).

To answer the question of whether the speakers of English as a second language are more likely to alternate languages than using English only when communicating misinformation online, we performed repeated measures analyses between LA_ST (mean $=$ 5.14, std $=$ 1.88) and EG_ST (mean $=$ 4, std $=$ 1.96). The results did not yield any significant difference ($p > .05$).

To answer the question about the relationship between the receiver type and language choice in the received misinformation online, we performed ANOVA by using the receiver type as the independent variable and RE_LA as the dependent variable. The results show that speakers of English as a second language (mean $=$ 4.36, std $=$ 2.02) are more likely to be exposed to online misinformation that involves language alternation than native English speakers (mean $=$ 2.20, std $=$ 1.86) ($p < .01$) (Table 1).

4.2 Qualitative Results

Two researchers analyzed the participants' responses to the open-ended questions independently using the open-coding method. Each of them created new categories based on

Table 1. Descriptive statistics (Mean [Standard Deviation]) of variables

Code	Variables	Speaker of English as a second language	Native English speaker
LA_ST	Language alternation	5.14 [1.875]	–
EG_ST	English use	4 [1.961]	–
RE_LA	Targeted misinformation w/language alternation	4.36 [2.023]	2.2 [1.859]

the themes emerged from analyzing the participants' textual responses. Finally, two sets of coding results were consolidated. We asked the participants why they would alternate languages in communicating misinformation online, and categorize their motivations as follows:

- Power of persuasion: Some participants felt that language alternation exerts the power of persuasion in online communication. For instance, "To achieve what I am looking for"; "Easy to convince that misinformation is true"; "It may make communicating misinformation more convincing"; and "Doing so takes less time to convince the other person".
- Universal language: Two participants responded that they alternated between their native language and English in online communication because English is more prevalent than other languages. For instance, "English is becoming a universal language and would look more appealing to most people." "English is understood by the majority of the mass and it is easier to convey information in that language."
- Image enhancing: Two participants perceived that alternating languages helps enhance their self-image. For instance, "make my appearance on dating websites look better."
- Belongingness: "Native language is used to make people comfortable and maybe win over them by showing belongingness to a particular sect."
- Genuineness pretending: Alternating languages helps them appear genuine and sincere to the communication target. For instance, "Create a sense of being genuine in the other person's eyes."
- Facilitating/hindering communication: On one hand, alternating languages makes the communication process easier, such as "easy to understand", "better communication", "to say my words exactly", and "get valuable information that this person knows". On the other hand, alternating languages contributes to "confusion" and "make the argument more complex."
- Self-protection: Several participants responded that alternating languages helps hiding information, identity, and feelings, such as "I want to hide my true feelings" and "Try to hide my identity."

We also asked the participants how they would respond if someone who communicates misinformation switched languages. We grouped their responses into the following categories.

- Suspicion: Several participants responded that alternating languages would arouse suspicion: "I wonder if it was sent by themselves", "I feel they are trying to hide something", "I am wary of them", and "Something's fishy". One participant even felt that "I see it as obvious."
- Confusion: Several participants felt confused when they saw someone's alternation between languages online.
- Curiosity and verification: A few participants were curious about what senders were up to and even performed fact checking as needed. For instance, "try to see what they're up to", "I will ask for clarification", and "I make sure to fact check."
- Acceptance: A few respondents did not consider it as uncommon. Instead, they felt "it is not a strange that someone is doing that." and "Depending on the individual, it would not be surprising if he/she is a stranger."
- Agitation: A handful of participants would feel anger and even agitated. "I would feel anger."
- Dismiss: A significant percentage of the participants indicated that they would choose to ignore the online misinformation once it is identified. For instance, "I won't reply to it." "I will ignore him (Just wasting my time)", "trash them", and "I would not care to the extent of trying to figure out whatever they are saying".

5 Discussion

The findings of this study reveal that language alternation does take place in communicating misinformation online. Speakers of English as a second language alternate languages as much as they use English in online misinformation communication. In addition, compared with native English speakers, it is more likely for speakers of English as a second language to be targeted by misinformation that involves language alternation.

The findings enrich the broader misinformation literature and add to the language alternation literature in a number of important ways. First, language alternation is a prevalent phenomenon in online communication, but it has been overlooked by misinformation studies to date. We propose several key questions regarding misinformation behavior in online communication and provide empirical evidence for language alternation in online misinformation communication. Second, we extend the language alternation theory by addressing online communication that involves misinformation. The current language alternation research is focused on face-to-face communication, but has rarely examined the context of online communication. Third, we address language alternation in online misinformation communication from both of the sender and receiver's perspectives. From a sender's perspective, the speakers of English as a second language are equally likely to use English only or alternate between English and their native languages. In addition, this study reveals important motives underlying language alternation in online misinformation communication for the first time. Interestingly, alternating languages can help not only enhance self-image and power of persuasion, but also facilitate genuineness pretending, self-protection, and strategic facilitation/hindrance of communication. These can be taken advantage by the sender of misinformation to increase their chance of success while lowering the possibility of arousing suspicion. From a receiver's perspective, native English speakers are less likely to be the target than non-native English speakers. The participants' reactions to online misinformation communication range from agitation through suspicion to acceptance. Some participants even planed for fact-checking out of curiosity to verify the information suspected of misinformation. Some other participants even did not consider language alternation in online misinformation communication as uncommon but treat it as reality that one has to face.

These findings have important implications for detecting online misinformation. There is a widespread recognition of the negative impacts of online misinformation on information credibility, but little attention to language alternation that keeps growing as the population of multilingual Internet users increases. Without knowing language alternation, it is difficult to understanding the scope of challenges in detecting online misinformation. This may result in poor performance in online misinformation detection and widespread diffusion of misinformation online.

By providing preliminary evidence that language alternation is not uncommon in communicating misinformation online, our findings can help improve misinformation detection by examining language alternation and offer insights into its prevalence and motives. The findings about the motives of using language alternation for communicating misinformation online elucidate its benefits, which can serve the purpose of misinformation communication through power of persuasion, self-protection, genuine pretense, and so on. These results extend the existing studies that have studied misinformation in a single language.

For information receivers, our findings demonstrate a wide range of reactions to language alternation in communicating misinformation online. Given the increasing globalization of online communication and collaboration, now is a particularly critical time to develop new approaches that have the potential for addressing language alternation in online communication, particularly for discovering distinctive patterns of language alternation in communicating misinformation compared with authentic information online.

We can carry on this important line of research in several directions. Although our empirical analyses and survey results suggest that language alternation exists in online misinformation communication, we did not analyze the actual online misinformation behavior. Another limitation is that the survey instruments used to measure the variables such as language alternation, English use, misinformation detection with language alternation, and targeted misinformation with language alternation can be further developed and validated. The focus of this study is on the language alternation behavior of speakers of English as a second language. There are many other speakers of English and non-English speakers who may alternate language in online communication. These can also be addressed in future research.

Acknowledgements. This research is partially supported by U.S. National Science Foundation (Award #: SES 1527684). Any opinions, findings, and conclusions expressed in this material are those of the authors and do not necessarily reflect the views of NSF.

References

1. Del Vicario, M., et al.: The spreading of misinformation online. Proc. Natl. Acad. Sci. **113**, 554–559 (2016)

2. Bessi, A., Coletto, M., Davidescu, G.A., Scala, A., Caldarelli, G., Quattrociocchi, W.: Science vs conspiracy: collective narratives in the age of misinformation. PloS ONE **10**(2), p. e0118093 (2015). https://doi.org/10.1371/journal.pone.0118093

3. Guess, A., Nyhan, B., Reifler, J.: Selective exposure to misinformation: evidence from the consumption of fake news during the 2016 US presidential campaign. Eur. Res. Counc. **9**(3), 1–49 (2018)

4. Grinberg, N., Joseph, K., Friedland, L., Swire-Thompson, B., Lazer, D.: Fake news on Twitter during the 2016 US presidential election. Science **363**(6425), 374–378 (2019)

5. Anderson, J., Rainie, L.: The future of truth and misinformation online. Pew Res. Cent. **19**, 1–224 (2017)

6. Hameleers, M., van der Meer, T.G.: Misinformation and polarization in a high-choice media environment: how effective are political fact-checkers? Commun. Res. **47**, 227–250 (2020)

7. Torabi Asr, F., Taboada, M.: Big data and quality data for fake news and misinformation detection. Big Data Soc. **6**(1), 1–14 (2019). https://doi.org/10.1177/2053951719843310

8. Zhang, C., Gupta, A., Kauten, C., Deokar, A.V., Qin, X.: Detecting fake news for reducing misinformation risks using analytics approaches. Eur. J. Oper. Res. **279**(3), 1036–1052 (2019). https://doi.org/10.1016/j.ejor.2019.06.022

9. Kumar, K.P.K., Geethakumari, G.: Detecting misinformation in online social networks using cognitive psychology. Human-centric Comput. Inf. Sci. **4**(1), 1–22 (2014). https://doi.org/10.1186/s13673-014-0014-x

10. Amorim, R.: Code switching in student-student interaction; functions and reasons! Linguística: Revista de Estudos Linguísticos da Universidade do Porto 7, pp. 177–195 (2012)

11. Bensen, H., Çavusoglu, Ç.: Reasons for the teachers' uses of code-switching in adult EFL classrooms. Hasan Ali Yücel Egitim Fakültesi Dergisi **10**, 69–82 (2013)

12. Eldin, A.A.T.S.: Socio linguistic study of code switching of the Arabic language speakers on social networking. Int. J. Engl. Linguist. **4**, 78–86 (2014)

13. Tuđman, M., Mikelic, N.: Information science: science about information misinformation and disinformation. In: Proceedings of Informing Science + Information Technology Education, vol. 3, pp. 1513–1527 (2003)

14. Vicario, M.D., Quattrociocchi, W., Scala, A., Zollo, F.: Polarization and fake news: Early warning of potential misinformation targets. ACM Trans. Web (TWEB) **13**, 1–22 (2019)

15. Shao, C., Hui, P.-M., Cui, P., Jiang, X., Peng, Y.: Tracking and characterizing the competition of fact checking and misinformation: case studies. IEEE Access **6**, 75327–75341 (2018)

16. Wang, P., Angarita, R., Renna, I.: Is this the era of misinformation yet: combining social bots and fake news to deceive the masses. In: Companion Proceedings of the The Web Conference 2018, pp. 1557–1561 (2018)

17. Kim, J., Tabibian, B., Oh, A., Schölkopf, B., Gomez-Rodriguez, M.: Leveraging the crowd to detect and reduce the spread of fake news and misinformation. In: Proceedings of the 11th ACM International Conference on Web Search and Data Mining, pp. 324–332 (2018)

18. Tschiatschek, S., Singla, A., Gomez Rodriguez, M., Merchant, A., Krause, A.: Fake news detection in social networks via crowd signals. In: Companion Proceedings of the The Web Conference 2018, pp. 517–524 (2018)

19. Woolard, K.A.: Codeswitching. In: A Companion to Linguistic Anthropology, pp. 73–94 (2004)

20. Gumperz, J.J.: Discourse Strategies. Vol. 1. Cambridge University Press, Cambridge (1982)

21. Halim, N.S., Maros, M.: The functions of code-switching in Facebook interactions. Procedia-Soc. Behav. Sci. **118**, 126–133 (2014)

22. Gammaldi, F.: Motivations for code-switching in blogs of Italian expats living in the Netherlands. Master's thesis, Universiteit Leiden, pp. 1–73 (2016)

23. Al-Qaysi, N., Al-Emran, M.: Code-switching usage in social media: a case study from Oman. Int. J. Inf. Technol. Lang. Stud. **1**, 25–38 (2017)
24. Jagero, N., Odongo, E.: Patterns and motivations of code switching among male and female in different ranks and age groups in Nairobi Kenya. Int. J. Linguist. **3**, 1–13 (2011)

A Taxonomy of User-Generated Content (UGC) Applications

Tien T. T. Nguyen[✉] and Arvind Tripathi

The University of Auckland, 12 Grafton Rd., Auckland 1010, New Zealand
`t.nguyen@auckland.ac.nz`

Abstract. User-generated content (UGC) has become well-known in the online context. Previous scholars have investigated UGC in terms of its helpfulness and utilization. Through the study of UGC and its applications, various practical implications including consumer behavior learning, IT system innovation, product search enhancement, user's value extraction, economic performance evaluation, and other applications have been ascertained and discussed. Until today, the applications of UGC are rapidly evolving. However, there has been little research into how UGC differently applied and operated. Also, little guidance is available for managers and researchers to recognize and compare the applications of UGC between and among the others. This study strives to fulfill the needs of further investigation on UGC applications regarding theory and conceptualization by developing a well-structured taxonomy of UGC applications. Because a taxonomy of UGC applications potentially enables the understanding of the science behind UGC. Eventually, understanding the insight of UGC provides the potential identification of the embedded UGC theories.

Keywords: UGC · User-generated content · UGC applications · Taxonomy · Classification

1 Introduction

During the past decade, the emergence of user-generated content (UGC) has marked a significant footprint in different disciplines, from business to science. UGC is regarded as a public good, is widely adopted and has become an integral and critical source of information for multiple stakeholders. UGCs are produced by nonprofessionals and mostly in a non-structured manner.

Despite a large body of research on UGC and its implications, no study has been conducted a particular classification for UGC or UGC applications. UGC is available in a large number of settings, including online trading, retailing, matchmaking, government, healthcare services, tourism and hospitality, sharing economy, learning and education, etc. Because of those vast amounts of available perspectives of UGC, proposing a representative taxonomy of UGC covering such a wide range of perspectives seems to be unlikely. Nevertheless, it is possible to classify the UGC applications. It is noteworthy that in this research, we define a UGC application from the usage perspective; e.g., to

© Springer Nature Switzerland AG 2020
K. R. Lang et al. (Eds.): WeB 2019, LNBIP 403, pp. 169–182, 2020.
https://doi.org/10.1007/978-3-030-67781-7_16

improve product search, estimate the trading risk, seek help/support in shopping, health-care, or during a disaster. Following that, each application in the paper is classified based on usage. More specifically, relying on the questions of "what" are the central dimensions/components of UGC applications? Also, "how" can UGC be adopted?, this study aimed to build a taxonomy of UGC applications where every entity/object of the application can be classified under critical principles. Each entity/object in the taxonomy is an application of UGC in a particular project. Moreover, each application is repre-sented by the main objective of each project. So, each target is a representative for each object/entity of interest in the taxonomy.

We adopted the methodology of taxonomy development suggested by Nickerson, Varshney, & Muntermann [1], which mapped out a validated process to develop a tax-onomy in information systems to build a taxonomy of UGC applications. The aims of the building of UGC applications are two-fold. Firstly, we seek an understanding of a set of specific concepts of UGC applications. From that, secondly, we organize physically the objects related to those concepts and make them possible to be differentiated, and easy to be recognized. The results of this study can serve as a theoretical foundation for future research. There are several practical implications as well. For instance, it enables a more focused, efficient research, and innovation of the utilization of UGC.

2   Theoretical Background

2.1   Taxonomy Definition

Taxonomy is defined as a particular science of classification from which an object of interest is classified by specific multi-dimensional characteristics that must satisfy the requirement of mutual exclusiveness and collective exhaustiveness. Following that, each object should be classified into one characteristic among many identified characteristics (a two or more) for each dimension; and no object can have less than or more than a single characteristic in each dimension. In brief, one and only one characteristic (Char) is allowed for each dimension (Dim) in a taxonomy (Taxo) as shown below:

$$Taxo = \left\{ Dim_i, i = 1 \ldots, n | Dim_i = \left\{ Char_{ij}, j = 1 \ldots, k_i, k_i \geq 2 \right\} \forall n \geq 2, Char_{ij} \subseteq Dim_i \right\}$$

2.2   Taxonomy of UGC Applications

The dramatic emergence of UGC and its applications in various contexts and domains ascertain its role. UGC dominates in most of the online marketplace platforms, specif-ically in terms of reputation systems (rating, feedback), customer reviews, customer comments, and Q&A. This study defines the UGC applications as the effectiveness of UGC in practices rather than the means of generating a content [2] where each application is regarded as an objective of a project utilizing UGC.

A taxonomy of UGC applications refers to a taxonomy of cognitive objects [3] because of the utilization of information (UGC) in suitable contexts or applications. In other words, it can be regarded as one of the classes of cognitive objects. The unit of analysis in this taxonomy development study is a UGC application. In summary, it is a classification from which every application of UGC is recognized and distinguished across specific dimensions and characteristics.

2.3 A Review of the Classification of UGC and UGC Applications

Information system (IS) literature presents extensive existence of taxonomies in different domains such as supply networks taxonomy [4], digital service design taxonomy [5], IS artifact evaluation methods taxonomy [6]. However, research is thin on UGC taxonomy or UGC application taxonomy with one exception [7]. The findings of Krumm and his colleagues [7] show that UGC and its applications are likely to be classified into four categories: data gathering, pattern recognition, community building, and public art. However, the classification of UGC of Krumm et al. [7] is preliminary and limited because it does not classify all types of UGC and UGC applications. Unlike the first and the second categories which are in connection with the usefulness of UGC, the third and fourth categories seem to be more of UGC motivations of contribution which belongs to a different meta-characteristic [1] of the UGC classification.

3 Methodology

To study the applications of UGC, we performed a systematic literature review to capture all possible applications. There have been 109 applications of UGC identified from 96 peer-reviewed journal papers. The selection of the documents is displayed in the following PRISMA diagram. Following that, a collection of 11,306 paper records were retrieved from the eight databases. The combination of multiple databases for the systematic literature review guarantees decent and efficient coverage [8]. Afterward, the dataset went through the next three phases of the selection process comprising screening, eligibility, and included to come up with the final data of 96 papers of analysis that need an in-depth review to study.

The study adopted the method of taxonomy development in IS proposed by Nickerson et al., [1] that followed seven iterative steps and tailored to the UGC application context to create the expected taxonomy. The process is described with the beginning of the meta-characteristics determination. Following is the decision of ending conditions. Then either an empirical to conceptual approach or a conceptual to empirical approach including steps four, five and six is determined to identify and examine the characteristics of the objects or entities. Finally, the decision-making stage is to terminate the process and produce the final taxonomy.

At first, the "central problem" of the taxonomy development process was specifying the meta-characteristics of the UGC application. As being a central point of projection which all the later bearing of characteristics shall be consistent with, the meta-characteristics should be comprehensive enough and capable of reflecting the primary purpose of the taxonomy. Besides, the identification of meta-characteristics also needs to engage with the users of the taxonomy. From the perspective of user's benefits, the choice of meta-characteristics can be further closed to the most brilliant features from which they make use of, resulting in the useful taxonomy.

Next, to identify the closing point for the following iterative approach of sub-characteristics identification after clarifying the meta-characteristics, ending conditions have been recognized. There were two types of ending conditions consisting of objective ending conditions, and subjective ending conditions originated from the study of Nickerson et al. [1] and validated in the context of UGC. Objective ending conditions were those satisfying the definition requirement of taxonomy. While subjective ending conditions are concerning the outcome evaluation of the researchers based on the standard criteria of concision, comprehensiveness, robustness, explanatory, and extendibility, objective ending conditions are those extracted from the definitions of taxonomy. A complete taxonomy is only set when all these identified conditions are met.

This study mostly counted on the inductive philosophical perspective in screening the dimensions and particular characteristics of the objects/entities, also regarded as an empirical to conceptual approach. Such an approach seems to be the best fit in this research because of available data sources of empirical studies from the review of literature in the domain of interest. Under the lens of pre-determined meta-characteristics, objects' characteristic was identified in succession. From which, the dimensions of the taxonomy were established. In this paper, we extracted different characteristics of UGC applications based on the literature review of many articles from a wide range of disciplines. The recognized characteristics were finally critically categorized into dimensions formulating an expected taxonomy. As mentioned, as all the proposed objective and subjective conditions were satisfied, the process of initial taxonomy development would complete.

Finally, we evaluated the resulting taxonomy. The work of evaluation was performed by approaching the conceptual to an empirical method which seeking for the appropriateness of the objects when classified into the resulting taxonomy. The process of assessment has thus initially confirmed the validation and usefulness of the taxonomy.

4   Development of a Taxonomy of UGC Applications

4.1   Identifying Meta-characteristics and Ending Conditions

The information quality of UGC and the appropriateness of UGC are most important for developing a taxonomy because they successfully characterize the effectiveness of each UGC application. Users of the UGC are consumers of information who will be likely to utilize the extracted information from the content dataset. Thus high-level characteristics of information which benefit the users are of importance. Since the meta-characteristics should be by the goals of the developed taxonomy, this study determines the meta-characteristics to be information quality and appropriateness of UGC applications.

Characterizing ending conditions is a crucial step before coming up with the iterative process of taxonomy development. Following Nickerson et al. [1], Table 1 demonstrates the ending conditions which we have used to examine and validate the final taxonomy.

4.2   Resulting Taxonomy

As the UGC application is the unit of analysis in this study, the adoption of UGC in solving a problem in various domains is an object or entity of this taxonomy development. Also, every entity represents every objective of UGC applications. Table 2 illustrates ten (out of 109) samples of applications of UGC equivalent to ten objectives accompanied by each brief description and sources of examples derived from the review of UGC literature.

Table 1. Proposed ending conditions for the development of the taxonomy (adopted from Nickerson et al., [1])

Objective ending conditions	The definition of a taxonomy satisfied	No object could have more than one characteristic in each dimension – Mutually exclusive restriction
		Each dimension must contain a characteristic associated with the object – Collectively exhaustive restriction
	Generalization achieved	All objects of interest or a representative sample of them have been investigated
	Comprehensiveness	Each characteristic of each dimension must include at least one classified object
		No changes (new, update, merge, split, or delete) of dimensions or characteristics in the last iteration.
		Every dimension is unique and within each dimension, every characteristic is unique
Subjective ending conditions	Concision	A parsimonious taxonomy is necessary with a limited number of dimensions and characteristics used
	Comprehensiveness	All the objects (applications) within the domain of UGC can be classified into this resulting taxonomy All the crucial dimensions of the objects should be fully identified
	Robustness	Sufficient dimensions and characteristics need obtaining to differentiate among objects
	Explanatory	All dimensions and characteristics are able to well explain the nature of every object

(continued)

Table 1. (*continued*)

	Extendibility	Additional dimensions or/and characteristics of new objects are allowed to extend the current taxonomy

Since the studied objects were recognized from previous research, the empirical to conceptual approach has been applied to develop the taxonomy. We identified the following characteristics and grouped them into appropriate dimensions.

Beneficiary: The first characteristics were related to the users from different domains such as business, R&D and innovation and social communities. We identify four types of users benefiting from applying UGC. These are - 1) buyers, 2) sellers, 3) researchers and practitioners; and 4) social support seekers. First, UGC offers valuable information for all aspects of a transaction including product, seller and platforms from diverse and experienced traders [9] for buyers. Second, UGC benefits sellers by building trust and reducing risk among anonymous traders [11, 25]. Third, researchers and innovators utilize UGC in innovating, designing and evaluating market mechanisms. Finally, UGC helps users seeking support from communities or crowdsourcing knowledge. These users can be members/followers of various online communities such as anticrime communities, fraud detection forums, health and beauty communities, virtual game communities, travelling blogs, and others. The extent and research (types of beneficiaries) of UGC applications vary in domain and time. In this case, we only consider the broader categories to be classified into the taxonomy.

Beneficiary = {Buyers (B), Sellers (S), Researchers and Practitioners (R&P), Social Support Seekers (SSS)}

UGC Source: Previous researches pointed out several forms of UGC from where users can obtain data and information. The characteristics related to UGC sources identified from various forms of UGC are text-based platforms, visual platforms or both. Text-based platforms are primary web pages or social media, blogs generating online reviews, feedback, comments. While visual platforms facilitate the generation of visual contents such as video clips, photos. The adoption of different types of UGC will match with its purpose of use, appropriateness of use, and availability of the database.

UGC Source = {Text − based Platforms (TP), Visual Platforms (VP), Multi−functional platforms (MFP)}

Multiplicity: Applications of UGC have engaged in the users' perspective of a group or an individual. The characteristic of a single adopter or a group of multiple adopters normally depends on the level of information and data analysis. At the basic utilization, an individual known as an executive of certain applications are always of interest. Meanwhile, only a group of users or an organization is able to optimize the usage of UGC to achieve information and data analysis at an advanced level. The recognition

Table 2. Identification of samples of UGC applications from the literature review

Application	Description	Research paper
Motivate/drive customers' intention to buy	UGC is utilized as a primary source of determinants of intention and decision to buy. UGC provides valuable information to enhance trust, reduce uncertainty. Customers use this information in their purchase decision	[9]
Customer behavior learning	UGC is utilized to investigate the purchase behavior. Also, UGC offers trends in trading engagement and involvement. Thus, analyzing those trends and involvement can discover consumer purchase behavior and identify consumer characteristics	[10, 11]
Improve the product searching process	The main objectives is to explore and improve the adoption of UGC in searching the products' characteristics	[16]
Online system innovation	UGC helps to examine the ability and innovation of online systems such as consumer ranking, a feedback mechanism or online review structure	[17, 18]
Evaluate economic performance	UGC enables the assessment of the economic performance of an individual or cooperation. Literature confirms the strong relationship between UGC and sales with regard to stock market performance, book sales, music sales, experience goods sales	[19]
Trading risk estimation & online fraud detection	Online communities can adopt UGC to learn about fraud and risks, especially in an online auction	[20]
Healthcare check-up support	Patients utilize UGC from online healthcare communities and social networks to follow up on their health conditions. They gain knowledge and help from the community	[21]
Predict business future profitability	UGCs are precious to enterprises when it to a specific degree show scenarios of business situations via the prediction of profit	[22]
Assessing the value creation	Through the interactions between stakeholders and clients extracted from the contents (UGC) especially in social media, value creation can be examined and evaluated	[23]

(continued)

Table 2. (*continued*)

Application	Description	Research paper
Support seeking in an extreme disaster	An emerging phenomenon since people tends to take advantage of contents left on the community to rescue the help seekers before, during and after a disaster	[24]

of these characteristics is associated with identified applications of UGC, specifically looking at the operators. We finally gathered these two characteristics and formulated the multiplicity dimension consisting of individual and group characteristics.

$$\text{Multiplicity} = \{\text{Individual (I)},\ \text{Group (G)}\}$$

Gratification: Literature [26] has identified the characteristics of content consumption of user-generated media (UGM) as *information fulfilling, entertainment, mood management needs, social connection,* and *online community enhancement.* We extend the adoption of these characteristics in a larger scope of UGC instead of UGM. It means, the first characteristic is defined as the *fulfillment of information* in all domains from business (e.g., sellers' information, product's features, customers' satisfaction), service (e.g., workmanship, service price), to social life (e.g., movies, healthcare, education), available on website, blogs or social media. Further, entertainment is not categorized as a separate characteristic because we are not looking at the dimensions from the view of entertainment but information utilization. Therefore, we consider *entertainment* is covered in fulfilling the information, and we use *pattern recognition* to replace instead, which has been identified as an important attribute [7]. We categorized these characteristics into the dimension, namely gratification. Finally, another scenario of information consumption is fostering innovation. This characteristic should be separated from the information fulfilling due to its specific utility; and because it is found as one of the popular objectives and implications in prior researches from literature reviews. All the identified characteristics and dimensions are logically and significantly correlated with the meta-characteristics.

$$\text{Gratification} = \{\text{Fulfilling Information (FI)},\ \text{Pattern Recognition (PR)},\ \text{Mood}$$
$$\text{Management Needs (MMN)},\ \text{Social connection enhancement (SCE)}\},\ \text{Fostering}$$
$$\text{innovation (FoI)}\}$$

Data Approach: The extraction of information from the text is crucial for the utilization of UGC but, the enormous volume, mishmash and unstructured feature of the data make it is hard to collect information from UGC otherwise to be possibly too subjective [17] and thus requires advanced technical support. The level of techniques and methods used are varied. It is sometimes only the fundamental extraction method such as semi-automated text mining tools to extract words or phrases of indicated concepts of product's feature, sellers' characteristics, sentiments attributes or linguistics styles from a large text library gathered in advance [16]. However, in some applications, a more skillful and advanced technology is needed to obtain the database and information. Irrespective of the technique, the resulting datasets should be more polished and precise than a manual approach. Therefore, we formulated data approaching dimensions consisting of the characteristics of raw data and processed data.

$$\text{Data Approach} = \{\text{Raw (R)}, \text{Processed (P)}\}$$

Operating Complexity: The idea of adoption maturity is based on four phases of technical skills [5] where the usefulness and appropriateness of UGC applications are found to be different. According to Williams et al., [5], development of four phases ranges from embedded system (no user requirement of technique skills), to consumer (might need little training on the advanced system of mobile applications), professional (users require to be able to learn more provided techniques to solve severe problems), and enthusiast (expertise skills are required to execute the operation and solve problems). Such stages are accompanied by four levels of skills: none, low, medium, and high respectively. We categorized the level of operating complexity into these phases of technical skills, which are supposed to be compatible in terms of maturity. A change in complexity should be followed by a suitable transformation at a technical skill level. Nevertheless, only low, medium and high skills are considered because applicators at least need several certain minimum skills to read and access the information they consume regardless of any level of complexity of the applications. So, we characterized low, medium beneficiaries and high as three attributes of the operating complexity dimension.

$$\text{Operating complexity} = \{\text{Low (L)}, \text{Medium (M)}, \text{High (H)}\}$$

False Information Existence: Quality of information is crucial for UGC applications, and the degree of quality varies in practices. It has a significant impact on the effectiveness of UGC adoption [27]. The existence of false information is one of the main reasons for low-quality information. Thus, the extent of false information in UGC captures information quality. Therefore, the characteristic of quality information was categorized via the identification of the extent of false information in the applications.

$$\text{False information existence} = \{\text{Low (L)}, \text{Medium (M)}, \text{High (H)}\}$$

Information Disclosure: The information and data in different applications of UGC are likely to vary with sources of retrieval and the extent of information disclosure. For example, low information disclosure is assigned for the sources, which are hard to retrieve information from due to privacy or very restricted access. We named this as private access. Alternatively, medium and high information disclosure were classified into the membership access characteristic and the open common access characteristic.

$$\text{Information disclosure} = \{\text{Open Common Access (OCA)}, \text{Membership Access}$$
$$\text{(MA)}, \text{Private Access (PA)}\}$$

4.3 Summary of the Taxonomy

As stated, the taxonomy will complete when all ending conditions are fulfilled. The last process of the taxonomy development provided satisfaction for all the identified characteristics and dimensions. Most of objective and subjective ending conditions are met. Briefly, Fig. 1 shows hierarchy layers of our completed taxonomy of UGC applications, and it is also constructed in the following formula (Fig. 2):

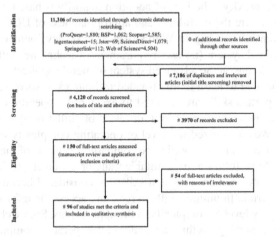

Fig. 1. The flow diagram of scholarly article selection

$\text{Taxo}_{UGC} = \{\text{Beneficial}(B,S,R\&P,SSS); \ \text{UGC\_Sources}(TP,VP,MFP);$
$\text{Multiplicity}(I,G); \ \text{Gratification}(FuI,PR,MMN,SCE,FoI); \ \text{Data\_Approach}(R,P);$
$\text{Operating\_Complexity}(L,M,H); \ \text{False\_Information\_Existence}(L,M,H);$
$\text{Information\_Disclosure}(OCA,MA,PA)\}$

5 Evaluation of the Resulting Taxonomy

We now evaluate the taxonomy of UGC applications by assessing its efficacy in classifying identified applications of UGC in the specific domain [1]. Briefly, the classification of these ten applications (a sample set of 109 identified applications) is fitted well into the current taxonomy. The result of the evaluation is presented below (Table 3):

6 Implications of the Taxonomy

The goal of this study is to create a useful taxonomy that is useful for intended purposes and intended users [1]. The users of the taxonomy of UGC applications were aimed to be researchers, managers, and practitioners. Also, the intended purpose was to serve

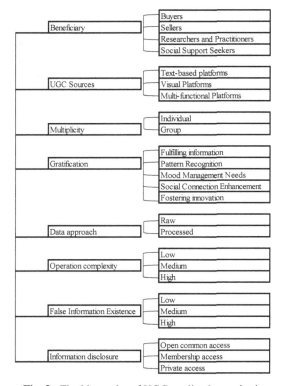

Fig. 2. Final herarchy of UGC applications criteria

as a descriptive and predictive instrument. From that, the taxonomy presented here has several potential uses. First, from the perspective of theorists/researchers/practitioners, who can apply this taxonomy to identify and differentiate their study from other similar works on UGC. Researchers/theorists can realize that they can resort to this taxonomy to conceptualize and define a particular usage of UGC in their studies. Given the set of identified characteristics, a distinct theme of a research topic relevant to that application of UGC is identified and described. This means bearing in mind the concept of the type and critical features of applications of UGC which are being investigated. And thence, the possible contribution to theory development for a better understanding of UGC and of how it is applied is obviously expected. Besides, our findings help practitioners concerning product, systems design and innovation. For example, designers can use the taxonomy to identify the characteristics associated with the usage of a UGC application. It could help designers of new projects that utilize UGC by identifying appropriate and important dimensions and characteristics. For example, if the target audience of the UGC application is users who are going to seek community help in a disaster, it should be associated with open access characteristic (first characteristic in the information disclosure dimension), which means, membership or user accounts may not be required.

Table 3. The classification of identified UGC applications into the taxonomy

	Benef icial	UGC Sources	Multipli city	Gratific ation	Data Approach	Operating Complexity	False Information Existence	Info. Disclosure
Motivate customers' intention to buy	B	TP	I	FuI	R	L	H	MA
Customer behavior learning	S	TP	G	PR	P	M	M	MA
Improve the product searching process	R&P	TP	G	FuI	P	H	L	OCA
Online system innovation	R&P	MFP	G	FuI	P	H	L	MA
Evaluate & predict economic performance	S	MFP	G	PR	P	M	M	MA
Trading risk estimation and online fraud detection	B	TP	I	FuI	P	M	H	OCA
Healthcare check-up support	SSS	TP	I	SCE	R	L	L	PA
Predict business future profitability	S	TP	G	PR	P	H	L	MA
Assessing the value creation	R&P	MFP	G	FoI	R	H	H	OCA
Support seeking in an extreme disaster	SSS	MFP	I	FI	R	L	H	MA

Second, from the perspective of project managers, who can assess, to some extent, the effectiveness of the project regarding UGC application dimensions. Managers are supposed to anticipate an intensive work of evaluating and managing their work/project effectively. Therefore, grasping the fundamentals of their specific project (an application of UGC) is essential. For instance, a manager of a project applying reviews of travelers to examine the determinants of overall satisfaction with their hotel in various travel group compositions. The taxonomy facilitates the identification of the work by pointing out the eight dimensions with correlative characteristics. So the manager may understand their project and from that instruct their project in a correct manner to save time and effort. If there is a high level of false information (a characteristic in false information existence dimension), it requires more effort from the data processor to better clean and detects the noise to avoid bias in the final results.

7 Future Work and Conclusion

This study has many limitations. The evidence for the accuracy and efficacy of the taxonomy is not compelling enough. Future work needs to explore the alternate method for evaluation to validate and assess the accuracy and efficacy of the resulting taxonomy instead of the manual analysis coding method. Thus, we will work on grouping the characteristics from the literature using statistical techniques such as clustering or factor analysis or conferring with experts in the relevant disciplines to improve the taxonomy evaluation. Such a robust taxonomy will allow us to develop a theoretical framework to explain and shape the objectives of UGC application projects. This is intended for future work.

In summary, in this paper, we created an initial taxonomy of UGC applications. The development of taxonomy comprised a continuous iterative process that adopted both conceptual-to-empirical and empirical-to-conceptual approaches to refine the critical characteristics and dimensions for a useful and comprehensive taxonomy. The study complied the technical instructions and standards in the methodology of taxonomy development in IS disciplines proposed by Nickerson et al., [1]. The results of the study are expected to provide rich insight into the understanding of UGC and its applications. Further, this paper strives to place a theoretical foundation for the classification of UGC. Thus, future researches can take advantage of this as a starting point for the continued improvement of the UGC concept and its relevant issues.

References

1. Nickerson, R.C., Varshney, U., Muntermann, J.: A method for taxonomy development and its application in information systems. Eur. J. Inf. Syst. **22**, 336–359 (2013). https://doi.org/10.1057/ejis.2012.26
2. Naab, T.K., Sehl, A.: Studies of user-generated content: a systematic review. Journalism **18**, 1256–1273 (2017). https://doi.org/10.1177/1464884916673557
3. Bloom, B.S.: Taxonomy of Educational Objectives. Cognitive Domain, vol. 1. McKay, New York (1956)
4. Harland, C.M., Lamming, R.C., Zheng, J., Johnsen, T.E.: A taxonomy of supply networks. J. Supply Chain Manag. **37**, 21–27 (2001). https://doi.org/10.1111/j.1745-493X.2001.tb00109.x
5. Williams, K., Chatterjee, S., Rossi, M.: Design of emerging digital services: a taxonomy. Eur. J. Inf. Syst. **17**, 505–517 (2008). https://doi.org/10.1057/ejis.2008.38
6. Prat, N., Comyn-Wattiau, I., Akoka, J.: A taxonomy of evaluation methods for information systems artifacts. J. Manag. Inf. Syst. **32**, 229–267 (2015). https://doi.org/10.1080/07421222.2015.1099390
7. Krumm, J., Davies, N., Narayanaswami, C.: User-generated content. IEEE Pervasive Comput. **7**, 10–11 (2008). https://doi.org/10.1109/MPRV.2008.85
8. Bramer, W.M., Rethlefsen, M.L., Kleijnen, J., Franco, O.H.: Optimal database combinations for literature searches in systematic reviews: a prospective exploratory study. Syst. Rev. **6**, 1–12 (2017). https://doi.org/10.1186/s13643-017-0644-y
9. Mudambi, S.M., Schuff, D.: What makes a helpful online review? A study of customer reviews on Amazon.com. MIS Q. **34**, 185–200 (2010). https://doi.org/10.2307/4132321
10. Gefen, D., Pavlou, P.A.: The boundaries of trust and risk: the quadratic moderating role of institutional structures. Inf. Syst. Res. **23**, 940–959 (2012). https://doi.org/10.1287/isre.1110.0395

11. Pavlou, P.A., Dimoka, A.: The nature and role of feedback text comments in online market-places: Implications for trust building, price premiums and seller differentiation. Inf. Syst. Res. **17**, 392–414 (2006). https://doi.org/10.1287/isre.1060.0106

12. Rudat, A., Buder, J.: Making retweeting social: the influence of content and context information on sharing news in Twitter. Comput. Human Behav. **46**, 75–84 (2015). https://doi.org/10.1016/j.chb.2015.01.005

13. Ye, Q., Law, R., Gu, B., Chen, W.: The influence of user-generated content on traveler behavior: An empirical investigation on the effects of e-word-of-mouth to hotel online bookings. Comput. Human Behav. **27**, 634–639 (2011). https://doi.org/10.1016/j.chb.2010.04.014

14. Jang, S., Prasad, A., Ratchford, B.T.: How consumers use product reviews in the purchase decision process. Mark. Lett. **23**, 825–838 (2012). https://doi.org/10.1007/s11002-012-9191-4

15. Huang, P., Lurie, N.H., Mitra, S.: Searching for experience on the web: an empirical examination of consumer behavior for search and experience goods. J. Mark. **73**, 55–69 (2009). https://doi.org/10.1509/jmkg.73.2.55

16. Han, W., McCabe, S., Wang, Y., Chong, A.Y.L.: Evaluating user-generated content in social media: an effective approach to encourage greater pro-environmental behavior in tourism? J. Sustain. Tour. **26**, 600–614 (2018). https://doi.org/10.1080/09669582.2017.1372442

17. Ghose, A., Ipeirotis, P.G., Li, B.: Designing ranking systems for hotels on travel search engines by mining user-generated and crowd-sourced content. Mark. Sci. **31**, 493–520 (2012). https://doi.org/10.2139/ssrn.1856558

18. Cha, M., Kwak, H., Rodriguez, P., Ahn, Y.Y., Moon, S.: I tube, you tube, everybody tubes: analyzing the world's largest user generated content video system. In: ACM SIGCOMM Conference, pp. 1–14 (2007). https://doi.org/10.1111/1756-185X.13112

19. Chevalier, J.A., Mayzlin, D.: The effect of word of mouth on sales: online book reviews. J. Mark. Res. **43**, 345–354 (2006). https://doi.org/10.1509/jmkr.43.3.345

20. Chua, H., Wareham, J., Robey, D.: The role of online trading communities in managing internet auction fraud. MIS Q. Manag. Inf. Syst. **31**, 759–781 (2007). https://doi.org/10.2307/25148819

21. Yan, L., Tan, Y.: Feel blue so go online: an empirical study of online supports among patients. Inf. Syst. Res. **25**, 690–709 (2014). https://doi.org/10.2139/ssrn.1697849

22. Tang, C., Mehl, M.R., Eastlick, M.A., He, W., Card, N.A.: A longitudinal exploration of the relations between electronic word-of-mouth indicators and firms' profitability: findings from the banking industry. Int. J. Inf. Manage. **36**, 1124–1132 (2016). https://doi.org/10.1016/j.ijinfomgt.2016.03.015

23. Suseno, Y., Laurell, C., Sick, N.: Assessing value creation in digital innovation ecosystems: a social media analytics approach. J. Strateg. Inf. Syst. **27**, 335–349 (2018). https://doi.org/10.1016/j.jsis.2018.09.004

24. Lu, Y., Yang, D.: Information exchange in virtual communities under extreme disaster conditions. Decis. Support Syst. **50**, 529–538 (2011). https://doi.org/10.1016/j.dss.2010.11.011

25. Masclet, D., Pénard, T.: Do reputation feedback systems really improve trust among anonymous traders? Exp. Study. Appl. Econ. **44**, 4553–4573 (2012). https://doi.org/10.1080/00036846.2011.591740

26. Shao, G.: Understanding the appeal of user-generated media: a uses and gratification perspective. Internet Res. **19**, 7–25 (2009). https://doi.org/10.1108/10662240910927795

27. Martins Nogueira, J.V.: The IQ of the crowd: understanding and improving information quality in structured user-generated content. Inf. Syst. Res. **25**, 669–689 (2014). https://doi.org/10.1017/CBO9781107415324.004

Global e-Business

Influence of Ownership and Management on IT Investment in Indian Family Firms

Xue Ning[1(✉)], Prasanna Karhade[2] ⓘ, Abhishek Kathuria[3] ⓘ, and Jiban Khuntia[1]

[1] University of Colorado Denver, Denver, CO 80204, USA
xue.ning@ucdenver.edu
[2] University of Hawaii at Manoa, Honolulu, HI 96822, USA
[3] Indian School of Business, Hyderabad, Telangana, India

Abstract. Several questions relevant to IT investments by family firms remain unanswered, such as, how family ownership influences a firm's IT investment and how the management type, including family management and professional management, further influences the family ownership and IT investment relationship. This study proposes several testable hypotheses, that can be investigated with data. A set of preliminary analysis using family firms' data from India suggest that family ownership is negatively related to IT investment. Furthermore, family management and professional management controls have different moderating effects on family ownership and IT investment relationship.

Keywords: IT investment · Family ownership · Family management · Professional management · India

1 Introduction

Family owned business constitute the majority of firms worldwide. Indian family businesses, with rich histories of domestic and international expansion (Celly et al. 2016), are distinguished from firms in other countries by a very high average level (about 50%) of family equity ownership and management (Hegde et al. 2016). There is a growing realization that G.R.E.A.T. domains (growing, rural, eastern, aspirational, and transitional), such as India, that account for a significant proportion of world population and economic output, warrant special in attention of practitioners and researchers alike (Karhade and Kathuria forthcoming). This is due to an important consideration - while management practices and lessons learned across the world can be extrapolated to local contexts, not all practices are universally applicable and realize equal benefits due to local nuances (Kathuria et al. 2018a, 2019, 2014). Thus, well-qualified professionals are in demand to manage family businesses and drive them to success. The tapping of a global pool of professionals has resulted in two distinct management controls, i.e., family and professionally managed firms. The extent to which these management controls are realized differs across firms, industries, and countries. This study aims to explore how these two management controls influence IT investment of family firms in India - an important emerging economy.

© Springer Nature Switzerland AG 2020
K. R. Lang et al. (Eds.): WeB 2019, LNBIP 403, pp. 185–193, 2020.
https://doi.org/10.1007/978-3-030-67781-7_17

Previous research reveals some preferences of family firms. For example, family ownership is negatively related to R&D investment, but this relationship becomes positive with moderating factors (Choi et al. 2015). Recent studies show that the way of management and governance influence the innovation of family firms (Matzler et al. 2015). Firm's heterogeneity, such as family CEOs' non-economic goals, influence the family firm's adoption of discontinuous technologies (Kammerlander and Ganter 2015). Studies state that investment decisions are complicated in family firms and show that family ownership negatively influences technology adoption based on the level of family influence (Souder et al. 2017). Family involvement also demonstrates the influence on the internationalization strategies of family firms (Fang et al. 2018; Ray et al. 2018).

This study asks three research questions: (1) whether family ownership influences IT investment, and if so, how? (2) How family management control moderates the impact of family ownership on IT investment? (3) How professional management control moderates the impact of family ownership on IT investment?

We merge the socioemotional wealth (SEW) perspective (Berrone et al. 2012) and organizational control theory (Hirschi and Gottfredson 1995; Ouchi 1979, 1980) as the theoretical framework of this study. SEW logic suggests that the family has a socioemotional connection with a firm, and the preservation of this socio-emotion has a role in strategic decision-making. However, management control would influence this socioemotion role—as suggested by the clan and bureaucratic control aspects of the control theory. Accordingly, control of the owning family, which is derived from the family's controlling position in a particular firm (Berrone et al. 2012) leads to the decisions about whether and not to invest in IT. We elaborate on this central premise of arguments to propose three sets of hypotheses and present a set of preliminary analysis using data collected from India.

2 Theory and Hypotheses

2.1 Ownership and the Socioemotional Wealth Perspective

The Socioemotional Wealth perspective has been used to explain why the strategic behavior of family-owned firms differs from nonfamily firms (Berrone et al. 2012; Gomez-Mejia et al. 2011, 2007). This perspective offers preservation of noneconomic or affective endowments of family owners as the primary motivation for the strategic behavior of the firm (Berrone et al. 2012).

The firm is a source of personal pride, self-identification, and satisfaction for the family (Schulze et al. 2001). Due to a resultant emotional attachment to the firm, family members usually frame strategic decisions as loss aversion of noneconomic endowments (Gomez-Mejia et al. 2007). In other words, the preservation and enhancement of socioemotional wealth is the primary strategy of family firms, which is reflected in their strategic decisions and outcomes (Berrone et al. 2012).

Due to the overarching desire to preserve socioemotional wealth, not all decisions made by family firms are economically justified. Family firms with undiversified ownership are risk-averse and demonstrate conservative strategic decision making. They avoid strategies that constitute a high risk and potentially high returns, such as internationalization (Ray et al. 2018) and R&D-intensive innovation (De Massis et al. 2013).

2.2 Management and Control Theory

While the nature of firm ownership influences the primary desire (preservation of socioe-motional wealth versus maximization of economic returns) underlying strategic deci-sions, owners must also possess the ability to exercise control over the firm's decisions and actions. Participation in the top management of a firm endows the ability to exercise control over the firm.

Organizational controls describe the primary mechanisms that organizations use to direct attention, motivation, and encourage organizational members to act in desired ways to meet an organization's objectives (Eisenhardt 1985; Ouchi 1979, 1980). Organizations use certain control mechanisms to ensure individual organizational members act in a manner that is consistent with achieving desired goals (Choudhury and Sabherwal 2003; Henderson and Lee 1992; Kirsch 1996; Kirsch et al. 2002; Kirsch 1997). Thus, control is the organization's attempt to increase the probability that organizational members will behave in ways that lead to the attainment of organizational goals (Henderson and Lee 1992).

2.3 Ownership and IT Investment

High family ownership discourages or negatively influences the adoption of a risky strategy such as high investment in IT because the family owners often bear the overall burden of risky investments and its detrimental impact on their reputation (in case of fail-ure of risky investments). Family ownership deters risky investments primarily because families are often motivated by not just by economic factors (Wright et al. 1996), but by broader socio-economic factors such as preservation of the long-term societal reputation of their businesses (Gomez-Mejia et al. 2011, 2007).

Family owners are particularly reluctant to invest in IT because of their preference for information asymmetry within the firm. Family owners prefer to operate by keeping information compartmentalized and within silos. This minimizes the risk of proprietary information from falling into the hands of competitors and thus harming the socioe-motional wealth of the family. Furthermore, the socioemotional wealth of the family is derived from not only the reputation and prestige of the firm but also by developing rela-tional capital by utilizing the influence of the firm (for example, by conferring favors). The presence of IT-driven business processes constrains family owners from leveraging the influence of the family firm for their benefit. Based on these, we propose:

Hypothesis 1: Family Ownership is negatively associated with IT Investment.

2.4 Management, Ownership, and IT Investment

A large proportion of family wealth is usually invested in a family firm, from which the family often derives the power and authority to impose the family's noneconomic goals on the firm (Carney 2005). Thus, a family member in the top management exerts clan control to fulfill the family's desire for conserving socioemotional wealth. In such firms, family managers have substantial control and decision-making authority, exerting a significant curtailing influence on the firm's IT investment strategy (Gomez-Mejia et al. 2011). Often, the confluence of ownership and management control provides the family with greater prestige and socioemotional wealth from the firm (Ray et al. 2018).

Further, family owners and family managers have similar risk aversion and aligned interests in the preservation and enhancement of socioemotional wealth and safeguarding of the family's name' for future generations (Anderson and Reeb 2003). Also, family owners and family managers have aligned interests in maintaining information asymmetry and leveraging informal business processes to exploit the influence of the firm towards generating relational capital for the family. This leads us to suggest:

Hypothesis 2: Family Management is likely to worsen the negative effect of Family Ownership on IT Investment.

Professional managers are hired because they possess knowledge and experience about successfully running similar businesses, have valuable business contacts, and other requisite skills and managerial capabilities lacked by the family. By hiring external managers, family owners cede some control over decision-making processes and these professional managers typically utilize bureaucratic control to achieve organizational goals in family-owned firms. The presence of top professional management may weaken the family owners' ability to exercise unconstrained authority, influence, and power over all aspects of the business (Schulze et al. 2001). This may help family owners recognize opportunities and best practices regarding IT investments and thus help family firms to overcome the aversion to IT investment (Miller et al. 2008).

Hypothesis 3: Professional Management is likely to weaken the negative effect of Family Ownership on IT investment.

3   Preliminary Analysis and Results

3.1   Setting

India, an important emerging economy, is the setting for our study. Recent studies (e.g., Kathuria et al. 2020; Kathuria et al. 2018a, b, 2016, 2019; Ramakrishnan et al. 2016) demonstrate that Indian context is helpful for scholars to understand the characteristics of firm strategies in a specific and important economy, that is unparalleled in its nuances. More specific to our study, family firms constitute a large percentage of publicly listed firms in India, and we used a sample of 2,148 firms, with 6,669 observations from 2006 to 2018 to conduct a set of preliminary analysis to suggest our hypotheses. Table 1 shows the variables used in our research.

Table 1. Variables table

Variable	Definition
Dependent variable	
Information Technology (IT) investment	Log of the total IT expenditures on software development charges, IT-enabled service charges, telephone, webhosting, satellite, internet, computer and IT systems, and software
Independent variables	
Family ownership share (FamOwn.)	The proportion of shares held by Indian individuals and Hindu undivided families as promoters (>20%)
Family management (FamMgt.)	The level of board management by family members. Coded as 1 if family member is a promoter and has the designation as Chairperson, CEO, or Managing Director. 0 otherwise
Professional management (ProfMgt.)	The level of board management by a non-family member (i.e., professional). Coded as 1 if non-family member is promoter and has the designation as Chairperson, CEO, or Managing Director. 0 otherwise
Control variables	
Firm age	The number of years of operation since firm's Incorporation year
Firm size	Log of the firm's sales
Liability	Log of the firm's total liabilities
R&D	Log of the firm's R&D expenditure
Marketing	Log of the firm's marketing expenditure
Prior performance	Percentage of earnings before depreciation, interest, and taxes to total assets
Year dummy	The year, within 2006–2018

3.2 Baseline Model

To test our hypotheses, we used the Generalized Least Squares (GLS) to estimate the generic model:

$$Y_{it} = X'_{it}\beta + \varepsilon_{it} \tag{1}$$

Where Y_{it} is dependent variable IT Investment, the X_{it} stands for explanatory variables, including family ownership, family management, professional management, the interaction of family ownership and family management, and the interaction of family ownership and professional management, the ε_{it} is the error term. Table 2 shows the main results of the GLS estimation analysis for the sample of 2,148 Indian family firms.

Model 1 is used to test our first hypothesis H1, which proposes the direct effect of family ownership on IT investment. As hypothesized, the result demonstrates that there is a significant and negative relationship between family ownership and IT investment (Table 2, column 1, $\beta = -0.004$, $p < 0.01$). In other words, the higher the family ownership of a family firm, the less IT investment. Therefore, hypothesis H1 is supported.

Table 2. Generalized least squares estimation results

	(1)	(2)	(3)	(4)
	Direct effect	Interaction1 FamMgt	Interaction2 ProfMgt	Both interactions
Variables	ITInvest	ITInvest	ITInvest	ITInvest
FamOwn.	−0.004***	−0.003***	−0.004***	−0.003***
	(0.001)	(0.001)	(0.001)	(0.001)
FamMgt.		0.353**		0.354**
		(0.141)		(0.141)
ProfMgt.			−1.999	−2.022
			(1.292)	(1.292)
FamOwn × FamMgt		−0.008***		−0.008***
		(0.003)		(0.003)
FamOwn × ProfMgt			0.079*	0.080*
			(0.043)	(0.043)
Control variables	Included	Included	Included	Included
Constant	−5.376***	−5.424***	−5.379***	−5.426***
	(0.075)	(0.077)	(0.075)	(0.077)
Observations	6,669	6,669	6,669	6,669
Number of firms	2,148	2,148	2,148	2,148

Note: GLS regression. Standard errors in parentheses *** $p < 0.01$, ** $p < 0.05$, * $p < 0.1$

We then test the moderating effects of family management on the direct relationship between family ownership and IT investment. Model 2 presents the analysis result for hypothesis H2 on the interaction effect of family ownership and family management. The result is significant and negative (Table 2, column 2, $\beta = -0.008$, $p < 0.01$), indicating the negative relationship between family ownership and IT investment become worse with the interaction of family management. This result is corresponding to H2 and thus supports H2. We examine the interaction of family ownership and professional management in column 3. The results suggest a significant and positive moderating effect of professional management on the relationship between family ownership and IT investment (Table 2, column 3, $\beta = 0.079$, $p < 0.1$), meaning professional management control could mitigate or weaken the negative impact of family ownership on the IT investment. As a result, hypothesis H3 is supported. Furthermore, the results in column

4 present the consistent moderating effects of family management and professional management.

4 Discussion

This study proposes set of hypotheses and conducts preliminary econometric analysis using data from India. We find that family management and professional management have opposite moderating effects on the relationship between family ownership and IT investment. First, we find the negative interaction between family ownership and family management. This result shows that when there is family management, the negative effect of family ownership on IT investment becomes worse. On the other hand, the interaction of family ownership and professional management is positive, indicating the adoption of professional management could increase a family firm's IT investment. These results imply that firms can use professional management controls to mitigate the negative effect of family ownership on IT investment but should control excessive family management.

This study contributes to both theory and practice. It provides several theoretical implications. First, we incorporate ownership and management control as two sources of heterogeneity among firms that explain differences in IT investment. Furthermore, we include control theory to discuss the potential role of management control as a reflection of ability over and above the primary desire asserted through the influence of firm ownership. Importantly, our findings shed light upon why IT intensity is relatively less in Indian firms (Kathuria et al. 2020; Kathuria and Konsynski 2012), and G.R.E.A.T. contexts as compared to developed nations (Karhade and Dong forthcoming; Karhade and Kathuria forthcoming).

This study also has important practical implications. First, it shows the negative impacts of family ownership on IT investment. In the modern competitive business environment, it is critical for firms to secure IT investment and maintain a competitive advantage. Therefore, if possible, a firm should try to mitigate the negative effects of family ownership to ensure IT investment. Second, for family firms, it might be difficult or impossible to decrease family ownership, management control could be another option to reduce the impact of ownership. For a firm that has high family ownership, this study suggests that professional management could be useful to mitigate the negative impact of high family ownership on IT investment.

In summary, this study suggests that the aversion of family firms towards IT investment is magnified when the family controls the firm, whereas professional managers can help reduce this aversion and encourage family-owned firms to make these necessary and critical future-oriented investments.

References

Anderson, R.C., Reeb, D.M.: Founding-family ownership and firm performance: evidence from the S&P 500. J. Finance **58**(3), 1301–1328 (2003)

Berrone, P., Cruz, C., Gomez-Mejia, L.R.: Socioemotional wealth in family firms: Theoretical dimensions, assessment approaches, and agenda for future research. Fam. Bus. Rev. **25**(3), 258–279 (2012)

Carney, M.: Corporate governance and competitive advantage in family–controlled firms. Entrep. Theory Pract. **29**(3), 249–265 (2005)

Celly, N., Kathuria, A., Subramanian, V.: Overview of Indian multinationals. In: Emerging Indian Multinationals: Strategic Players in a Multipolar World. Oxford University Press, London (2016)

Choi, Y.R., Zahra, S.A., Yoshikawa, T., Han, B.H.: Family ownership and R&D investment: the role of growth opportunities and business group membership. J. Bus. Res. **68**(5), 1053–1061 (2015)

Choudhury, V., Sabherwal, R.: Portfolios of control in outsourced software development projects. Inf. Syst. Res. **14**(3), 291–314 (2015). 2003

De Massis, A., Frattini, F., Lichtenthaler, U.: Research on technological innovation in family firms: Present debates and future directions. Family Bus. Rev. **26**(1), 10–31 (2013)

Eisenhardt, K.M.: Control: organizational and economic approaches. Manage. Sci. **31**(2), 134–149 (1985)

Fang, H., Kotlar, J., Memili, E., Chrisman, J.J., De Massis, A.: The pursuit of international opportunities in family firms: Generational differences and the role of knowledge-based resources. Glob. Strategy J. **8**(1), 136–157 (2018)

Gomez-Mejia, L.R., Cruz, C., Berrone, P., De Castro, J.: The bind that ties: socioemotional wealth preservation in family firms. Acad. Manage. Ann. **5**(1), 653–707 (2011)

Gomez-Mejia, L.R., Haynes, K.T., Núñez-Nickel, M., Jacobson, K.J., Moyano-Fuentes, J.: Socioemotional wealth and business risks in family-controlled firms: Evidence from Spanish olive oil mills. Admi. Sci. Q. **52**(1), 106–137 (2007)

Hegde, S.P., Seth, R., Ramanna, V.: Are Shareholders of Family Firms Really Better Off? Working Paper (2016)

Henderson, J.C., Lee, S.: Managing I/S design teams: a control theories perspective. Manage. Sci. **38**(6), 757–777 (1992)

Hirschi, T., Gottfredson, M.R.: Control theory and the life-course perspective. Stud. Crime Crime Prev. **4**(2), 131–142 (1995)

Kammerlander, N., Ganter, M.: An attention-based view of family firm adaptation to discontinuous technological change: exploring the role of family CEOs' noneconomic goals. J. Prod. Innov. Manage **32**(3), 361–383 (2015)

Karhade, P. P., and Dong, J. Q.: Innovation outcomes of digitally enabled collaborative problemistic search capability. MIS Q. (forthcoming)

Karhade, P.P., Kathuria, A.: Missing Impact of Ratings on Platform Participation in India: A Call for Research in GREAT Domains. Communications of the Association for Information Systems (forthcoming)

Kathuria, A., Karhade, P.P., Konsynski, B.R.: In the realm of hungry ghosts: multi-level theory for supplier participation on digital platforms. J. Manage. Inf. Syst. (forthcoming)

Kathuria, R., Kathuria, N.N., Kathuria, A.: Mutually supportive or trade-offs: an analysis of competitive priorities in the emerging economy of India. J. High Technol. Manage. Res. **29**(2), 227–236 (2018a)

Kathuria, A., Konsynski, B.R.: Juggling paradoxical strategies: the emergent role of IT capabilities. In: Proceedings of the International Conference on Information Systems. Association of Information Systems, Orlando (2012)

Kathuria, A., Mann, A., Khuntia, J., Saldanha, T.J.V., Kauffman, R.J.: A strategic value appropriation path for cloud computing. J. Manage. Inf. Syst. **35**(3), 740–775 (2018b)

Kathuria, A., Saldanha, T.J.V., Khuntia, J., Andrade Rojas, M.G.: How information management capability affects innovation capability and firm performance under turbulence: evidence from India. In: Proceedings of the International Conference on Information Systems. Association of Information Systems, Dublin (2016)

Khuntia, J., Saldanha, T.J.V., Kathuria, A.: Dancing in the Tigers' Den: MNCs versus local firms leveraging IT-enabled strategic flexibility. In: Proceedings of the International Conference on Information Systems. Association of Information Systems, Auckland (2014)

Khuntia, J., Saldanha, T.J.V., Kathuria, A., Konsynski, B.R.: Benefits of IT-enabled flexibilities for foreign versus local firms in emerging economies. J. Manage. Inf. Syst. **36**(3), 855–892 (2019)

Kirsch, L.J.: The management of complex tasks in organizations: controlling the systems development process. Organ. Sci. **7**(1), 1–21 (1996)

Kirsch, L.J., Sambamurthy, V., Ko, D.-G., Purvis, R.L.: Controlling information systems development projects: the view from the client. Manage. Sci. **48**(4), 484–498 (2002)

Kirsch, L.S.: Portfolios of control modes and IS project management. Inf. Syst. Res. **8**(3), 215–239 (1997)

Matzler, K., Veider, V., Hautz, J., Stadler, C.: The impact of family ownership, management, and governance on innovation. J. Prod. Innov. Manage **32**(3), 319–333 (2015)

Miller, D., Le Breton-Miller, I., Scholnick, B.: Stewardship vs. stagnation: an empirical comparison of small family and non-family businesses. J. Manage. Stud. **45**(1), 51–78 (2008)

Ouchi, W.G.: A conceptual framework for the design of organizational control mechanisms. In: Emmanuel, C., Otley, D., Merchant, K. (eds.) Readings in Accounting for Management Control, pp. 63–82. Springer, Bosto (1979). https://doi.org/10.1007/978-1-4899-7138-8_4

Ouchi, W.G.: Markets, bureaucracies, and clans. Adm. Sci. Q. **25**(1), 129–141 (1980)

Ramakrishnan, T., Khuntia, J., Kathuria, A., Saldanha, T.J.V.: Business intelligence capabilities and effectiveness: An integrative model. In: Proceedings of the 49th Hawaii International Conference on System Sciences. IEEE, Koloa (2016)

Ray, S., Mondal, A., Ramachandran, K.: How does family involvement affect a firm's internationalization? An investigation of Indian family firms. Glob. Strategy J. **8**(1), 73–105 (2018)

Schulze, W.S., Lubatkin, M.H., Dino, R.N., Buchholtz, A.K.: Agency relationships in family firms: theory and evidence. Organ. Sci. **12**(2), 99–116 (2001)

Souder, D., Zaheer, A., Sapienza, H., Ranucci, R.: How family influence, socioemotional wealth, and competitive conditions shape new technology adoption. Strateg. Manage. J. **38**(9), 1774–1790 (2017)

Wright, P., Ferris, S.P., Sarin, A., Awasthi, V.: Impact of corporate insider, blockholder, and institutional equity ownership on firm risk taking. Acad. Manage. J. **39**(2), 441–458 (1996)

Controlling Risk from Design Changes in Chinese Prefabricated Construction Projects: An Empirical Investigation

Juan Du[1,3(✉)], Jiajun Zhang[1], Yifei Gu[1], and Vijayan Sugumaran[2]

[1] SILC Business School, Shanghai University, Shanghai, China
ritadu@shu.edu.cn
[2] Department of Decision and Information Sciences, School of Business Administration, Oakland University, Rocheste, USA
[3] Faculty of Engineering and IT, University of Technology Sydney, Sydney, Australia

Abstract. Industrialization and standardization of prefabricated construction has brought reform and improvements to the construction industry and has changed the traditional construction mode. Prefabricated construction reduces labor demand and environmental pollution, which is in response to the government's "low-carbon environmental protection, green industry" call. Prefabricated projects, while improving engineering quality and work efficiency, also entail complexity and uncertainty. The development of industrialization projects in China still lags behind, and many risks restrict the development of successful prefabricated projects. This paper focuses on the specific problem of design change risk, and uses an empirical study to study the impact of design change risk factors, identify and prioritize them for different design change risk events. We also propose specific management strategies accordingly.

Keywords: Prefabricated construction · Risk control · Empirical investigation · SEM

1 Introduction

Compared to traditional onsite construction, prefabricated construction uses components produced in an external factory and transported to the construction site for assembly. Prefabricated components are mass produced in industrial assembly lines (Qi and Li 2014). Thus, prefabricated construction has significant advantages with respect to environmental protection, efficiency and resource saving. This is very much in line with the goals of green industry and sustainable development.

While bringing about positive changes, prefabricated construction is also accompanied by many complexities and uncertainties in project management (Yang and Wang 2008). Understanding the risk factor is the main task of risk control (Layth et al. 2019). The design change risk investigated in this paper belongs to engineering change risk, which is the most common change risk event in assembled construction projects (Du and

© Springer Nature Switzerland AG 2020
K. R. Lang et al. (Eds.): WeB 2019, LNBIP 403, pp. 194–214, 2020.
https://doi.org/10.1007/978-3-030-67781-7_18

Jing 2019). The need to accommodate design change will lead to delay project schedule, increase cost, disrupt normal working steps and increase the complexity of the project.

In view of the characteristics and main problems associated with design change risk in assembled buildings, this paper explores the design change risk events in prefabricated construction projects and identifies the corresponding influencing factors. Finally, according to the priority of the factors and project characteristics, corresponding measures are designed for reasonable control.

2   Research Background

2.1   Study on Risk Control of Prefabricated Projects

There are some existing studies investigating the risks associated with prefabricated projects and ways to control them. The design system of prefabricated assembly building based on knowledge based system was first proposed by Retik and Abraham (1994). They assemble all aspects of the design system, including architecture, technology, technical scheme and structure, into a module. First, they make a preliminary design, then adjust the structure of the module, and then divide it into prefabricated components. Finally, they give a rough cost estimate and specific drawings of prefabricated components. To some extent, this system provides corresponding strategies for cost risk and engineering change risk in assembly building. Wang et al. (2018) describe a network model of project objectives, risk events, risk factors and stakeholders in the process of building industrialization based on meta-network analysis. The main process of risk management in the model consists of three parts: risk identification, risk analysis & assessment, and risk treatment & control. Risk factors are identified according to literature and field investigation, and indirect influences on project objectives are calculated by network analysis and algorithm calculation in metanetwork. Key risk factors are analyzed to determine the stakeholders that affect these factors, so as to propose relevant control strategies. Li Sihuan et al. proposed a multi-agent-based information sharing coordination mechanism for the supply chain, providing a centralized information sharing platform for various agents to collect data to be exchanged. Zhang Xian studied the integrated management scheme of construction project based on agent method. From the perspective of complex adaptability, the project management maturity theory is applied to construct the simulation optimization model of construction project system, and some related factors that restrict the maturity of project management are found out. This helps enterprises to carry out better strategic adjustment and management of construction projects.

2.2   Empirical Method in Engineering Management

For the risk control mechanism of prefabricated assembly projects, the research methods adopted in this paper are survey questionnaire and structural equation model (Sheng-Hua 2006, Tong 2007, Yue et al. 2010, Ali and Gholamreza 2019). The purpose of this paper is to formulate strategies for key inducers; thus, the key inducers and key stakeholders should be investigated through the questionnaire. The basic elements of the questionnaire include project objectives (cost and duration), design change risk events, uncertainties, and stakeholders.

3 The Characterization of the Problem

This paper puts forward management strategies based on two aspects of project character-istics and risk factor priority. Risk factors are analyzed using structural equation model, and the data used in structural equation model are collected through questionnaire.

3.1 Risk Elements and Risk Events

Risk events of design change will occur in prefabricated construction, and a large num-ber of stakeholders will be involved, leading to losses between partners. Based on the literature and industry experiences, there are four major categories of design change risk events. The first category is the change of investor (owner) demand, the second category is the design change in the deepening design, the third category is the design change in the production or assembly, and the fourth category is the design change in the coordi-nation management. Therefore, it is necessary to identify all the uncertainty factors that may cause design changes, and manage these triggering factors to achieve reasonable and effective management control of design changes. In this paper, 11 common and representative risk events were extracted and analyzed through a questionnaire survey (Risk1–Risk11):

R1. Due to market change or capital adjustment, the owner proposed to reduce the amount of project. The original plan was to build 20 floors, but due to the lack of capital turnover, now it is only 15 floors.

R2. The owner adjusts the building's function and proposes to change the location of the embedded parts.

R3. Unreasonable design scheme, missing embedded parts and cable boxes, etc., need to be changed.

R4. The designer drew the dimensions of reserved holes of components and grouting sleeve wrong, and made design changes through feedback from the manufacturer.

R5. Improper coordination of the designers' specialties results in collision between the reserved orifice and the steel bar.

R6. The designer does not understand the process, which leads to the failure of mold release in the production of wallboard components, which needs to be changed.

R7. Too many orders from the manufacturer led to the production delay of components, and the prefabricated components were returned by the installation side and changed to cast-in-situ.

R8. Collision between steel bar and reserved opening results in change.

R9. The composite plate is cut due to the failure to consider the size of the prefabricate components when the steel bar is arranged by the assembly side.

R10. Improper coordination between the installation party and the design and produc-tion party, resulting in waste of components, difficulties in field connection and other problems.

R11. The information is not updated, resulting in inconsistency between PC component drawings and construction drawings.

Some change events are caused by a single trigger, while others are caused by a combination of multiple triggers. The prefabricated construction project supply chain is a very complex engineering system. It is discrete, dynamic and composed of multiple collaborative participation. And the stakeholders have great differences in project experience and expertise, which leads to the non-interoperability of information and the discrete distribution of information in each link, making the management decision-making very complex.

Based on the literature and expert interviews, the following analysis is made on the factors inducing design risk change in prefabricated projects:

Table 1. Design change risk inducers.

Factor type	Segmentation of risk factors	Number
Integrated management capability		
Personnel management (A)	Inadequate planning, control and problem-solving capacity of project participants	A1
	Participants did not pay enough attention to the project plan and did not have a correct attitude	A2
	Lower level of knowledge of staff	A3
Resource management (B)	The design of income distribution mechanism is not scientific enough	B1
	Inappropriate time allocation by project managers	B2
	Insufficient funds for project participants	B3
Organizational management (C)	Flexibility of Organizational Structure	C1
	Imperfect system of standards and norms	C2
	Conflict of management perceptions among organizations	C3
Partnerships		
Information and communication (D)	Low level of information sharing among project participants	D1
	Design experience of similar projects in design unit or design team	D2
	Insufficient experience of component factory or construction personnel	D3
	Low frequency of communication between project participants and teams	D4

(*continued*)

Table 1. (*continued*)

Factor type	Segmentation of risk factors	Number
Collaboration mechanism (E)	Poor information feedback and coordination among parties	E1
	The construction unit and the production unit shall not participate in the design	E2
	The design team is not involved in component production and construction	E3
	Unreasonable design of partner selection mechanism	E4
	Partner risk sharing is low	E5
	Poor sharing of resources among partners	E6
Technical application		
Technical level (F)	Lack of technological innovation capacity	F1
	Low matching of technical solutions	F2
	The technical level is low	F3
	Component Splitting and Deepening Design Technology Immature	F4
	The production efficiency of component factory is low	F5

The Table 1 above lists a number of controllable risk factors for design changes, provides variables A–F for the construction of conceptual model. And each major factor contains 3 to 6 minor factors. At the same time there are some uncontrolled factors that can lead to design changes, such as force majeure factors leading to the suspension of the project, or the macroeconomic impact, for example, national reorganization.

This paper proposes management strategies with respect to two aspects of project characteristics, which are based on the correlation between risk factors and risk results, and the priority of risk factors.

Table 2. Correlation between risk factors and risk outcomes.

Risk factor	Associated risk event
A1	R7, R10
A2	R3–R9
A3	R4, R6, R8
B1	R1

(*continued*)

Table 2. (*continued*)

Risk factor	Associated risk event
B2	R3, R4, R7
B3	R1
C1	R5–R7, R10, R11
C2	R3–R11
C3	R6, R10
D1	R2, R9, R10, R11
D2	R2, R9, R10, R11
D3	R3, R4, R5, R6
D4	R8, R9
E1	R7, R10, R11
E2	R4, R6
E3	R10, R11
E4	R4–R6;R10
E5	R4;R6–R10
E6	R7, R9
F1	R8, R9
F2	R6
F3	R7–R9
F4	R4–R6;R9
F5	R7

The Table 2 above shows the risk events associated with uncertain factors. For example, when there is factor A1: the planning arrangement of project participants is not in place, R7 and R10 risk events are likely to occur. Different management strategies need to be developed according to these risk factors and different project characteristics.

3.2 Empirical Method in Engineering Management

Based on the existing literature and interview of industry experts, six main controllable inducement factors of design change risk of prefabricated construction were identified: Personnel management ability (A); Resource allocation capacity (B); Organizational capacity (C); Information and experience among stakeholders (D); Mechanisms for collaboration among stakeholders (E); Technology applications (F). According to interviews with experts, it is found that these six factors all have an impact on the design risk of prefabricated construction. Therefore, based on the risk factors, the following conceptual model can be constructed:

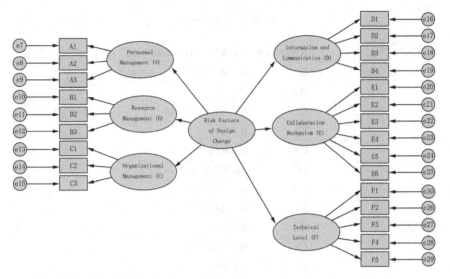

Fig. 1. Model of risk factors impacting prefabricated construction design change.

The theoretical model is analyzed using Amos21 software. As shown in Fig. 1, the research model contains 6 latent variables: A–F and design change risk, and 24 observed variables: A1–F5.

After the establishment of the theoretical model, it is necessary to make assumptions about the path relationships in the model. According to the causal relationship among various variables and the results of expert interviews, the following hypotheses are proposed:

H1: The personnel management factor plays a positive role in promoting the risk of controlling the change of prefabricated construction design.

H2: Resource allocation factors play a positive role in promoting the risk control of prefabricated construction design changes.

H3: Organizing production factors plays a positive role in controlling the risk of change of prefabricated construction design.

H4: Information and experience among partners play a positive role in controlling the risk of design changes of prefabricated construction.

H5: The cooperative mechanism among the partners has a positive influence on the risk control of prefabricated construction design change.

H6: The factors of technology application level play a positive role in promoting the risk control of prefabricated construction design change.

4   Empirical Analysis

4.1   Data Collection

Data needed by the model were collected through survey questionnaire, and a total of 103 questionnaires were collected. The middle and senior managers of related enterprises participated in the questionnaire survey, and the questionnaire recovery rate was 97%. The commonly used methods of questionnaire data test are reliability test and validity test. The purpose of reliability test is to detect the consistency and stability of data. The higher the reliability test coefficient is, the lower the error is, and the data is more practical. Validity test is used to ensure the accuracy and validity of data.

Reliability Test
The most commonly used method for reliability test is to calculate Cronbach's Alpha. The value of Cronbach's Alpha is within the range of $(0,1)$, and the closer the coefficient value is to 1, the higher the reliability.

Table 3. Reliability test of potential variables.

Latent variable	Number of items	Cronbach's Alpha
Personnel management factors (A)	3	0.898
Resource allocation factor (B)	3	0.895
Organizational capacity (C)	3	0.878
Information and Experience Factors (D)	4	0.952
Collaboration mechanism factor (E)	6	0.951
Technical scope factors (F)	4	0.723
Overall	23	0.923

Based on the test, the reliability index of F1 is relatively low. After F1 is deleted, the test results are shown in the Table 3 above, and the reliability index is good.

Validity Test
Validity is generally measured by KMO and Bartlett sphericity test. SPSS was used for validity analysis. The test results and overall results of each variable are given in the following Table 4:

Table 4. KMO and Bartlett sphericity test.

KMO and Bartlett's test		
KMO Number of sample suitability measures		0.823
Bartlett sphericity test	Approximate chi-square	2057.536
	Degree of freedom	253
	Significance	0

As can be seen from the above table, the overall KMO value of all variables is 0.823, greater than 0.7, indicating that there is a good correlation between variable data and research issues. In bartlett's sphericity test, the approximate card was 2057.536, the degree of freedom was 253, and the significance was 0.000 and significantly less than 0.001, which verified that the questionnaire had reasonable structure. KMO and bartlett test show that the data collected using this questionnaire is suitable for factor analysis.

Thus, based on the reliability and validity test conducted using SPSS software, the results show that the questionnaire has reliable validity and stability, and the structure of the questionnaire is sound, and the content correlation is also at normal levels. Hence, the data collected using this survey questionnaire is effective and reliable.

4.2   Data Collection

The proposed research model is used to verify whether the inducer can effectively control the risk of design change and propose the corresponding decision plan according to the priority of each factor.

According to the previously proposed prefabricated building project design risk factor theory model, create the latent variables and observed variables of the theoretical model in Amos, and make appropriate adjustments according to the results of data inspection. The latent variables are represented by A–F, observation variables were expressed as A1-3, B1-3, C1-3, D1-4, E1-6, and F2-5, as shown in the figure below (Fig. 2):

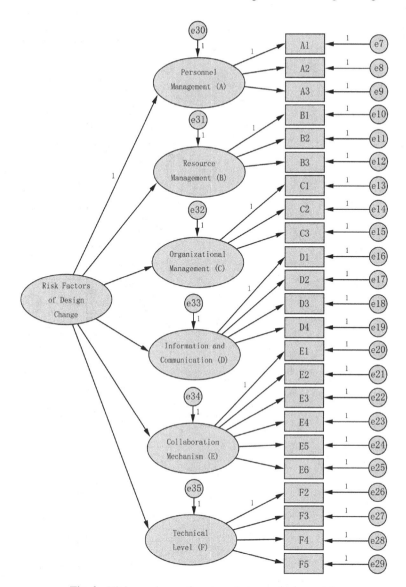

Fig. 2. Higher-order confirmatory factor analysis model.

On the basis of the proposed theoretical model, F1 variable is removed according to the results of the reliability of data. Model fitting results are shown in Fig. 3. Specific fitting index is shown in the following Table 5:

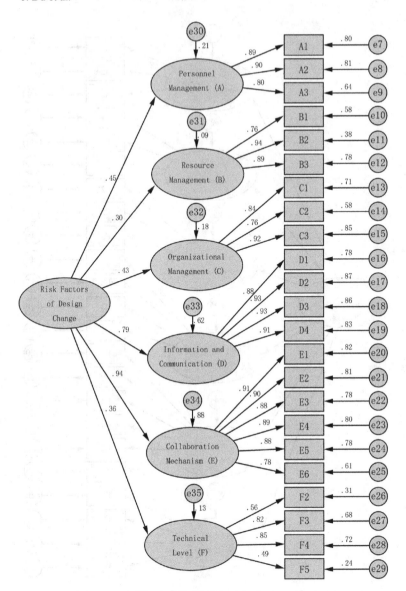

Fig. 3. Model fitting results.

According to the coefficient analysis results of the model fitting index, the chi-square degree of freedom ratio (CMIN/DF) is 1.890, which is within the range of the ideal standard (1,3). This index represents a good model fit. The other indicators are basically at a good level, and some of them reach a general or bad state. In general, the model fitting condition is good, and the fitness between the actual data and the theoretical model is very good, and the model results are convincing.

Table 5. Model fitting index coefficient analysis.

Fitness test index	Ideal standard	General standard	Model results	Conclusion
CMIN/DF	(1,3)	The smaller the better	1.89	Good
RMSEA	<0.08	<0.1	0.093	General
RMR I RMR	<0.08	<0.1	0.094	General
GFI	>0.90	>0.8	0.748	Bad
CFI	>0.90	>0.8	0.9	Good
IFI	>0.90	>0.8	0.901	Good
PNFI	>0.50		0.718	Good

In order to further refine operation results and analyze confirmatory factors, detailed parameter estimation is shown in Table 6.

Table 6. Parameter estimation of confirmatory factor.

Variable			Estimate	Variable			Estimate
A1	<---	A	0.894	D4	<---	D	0.913
A2	<---	A	0.9	E1	<---	E	0.906
A3	<---	A	0.8	E2	<---	E	0.899
B1	<---	B	0.764	E3	<---	E	0.881
B2	<---	B	0.936	E4	<---	E	0.893
B3	<---	B	0.886	E5	<---	E	0.883
C1	<---	C	0.843	E6	<---	E	0.784
C2	<---	C	0.764	F2	<---	F	0.555
C3	<---	C	0.921	F3	<---	F	0.823
D1	<---	D	0.881	F4	<---	F	0.848
D2	<---	D	0.931	F5	<---	F	0.488
D3	<---	D	0.927				

Based on the path coefficient index analysis, we believe that in the measurement model, the observation variable can well reflect the latent variable when the path coefficient is greater than 0.6, and retain the variable. When the observed variables cannot effectively reflect latent variables, they need to be deleted. Based on the above standards, it is found that the model still has room for adjustment and needs to be modified.

4.3 Model Modification

Referring to the coefficient standard proposed in the previous part, in the observation model, the index whose path coefficient is less than or equal to 0.6 in the observation variable is deleted, namely F2 (0.555) and F5 (0.488) are deleted, and the model is further modified according to the MI index after the first revision. The variance analysis after the first revision of the model is shown in the following Table 7:

Table 7. Variance estimation.

			M.I.	Par change
e27	<-->	e30	5.638	0.077
e24	<-->	e30	20.028	0.217
e24	<-->	e25	4.305	0.071
e23	<-->	e32	5.984	0.105
e23	<-->	e24	5.951	0.062
e22	<-->	e32	5.227	−0.11
e22	<-->	e23	4.136	−0.054
e21	<-->	e32	4.873	−0.097
e21	<-->	e30	8.126	−0.133
e21	<-->	e24	9.541	−0.079
e21	<-->	e22	13.801	0.101
e20	<-->	e27	6.897	0.044
e16	<-->	e25	5.089	−0.075
e15	<-->	e24	4.933	0.074
e15	<-->	e16	5.094	0.072
e14	<-->	e36	8.261	0.101
e13	<-->	e28	6.4	0.061
e13	<-->	e18	8.49	0.093
e13	<-->	e17	5.074	−0.07
e13	<-->	e16	7.16	−0.088
e12	<-->	e15	10.257	−0.118
e12	<-->	e13	4.173	0.078
e11	<-->	e23	4.46	0.057
e11	<-->	e15	4.396	0.073
e10	<-->	e24	4.729	0.081
e10	<-->	e20	9.859	−0.109
e10	<-->	e19	8.036	−0.103
e9	<-->	e26	6.372	−0.064
e9	<-->	e25	6.672	0.124
e9	<-->	e10	4.561	0.111
e8	<-->	e33	7.895	−0.104
e8	<-->	e27	11.433	0.071

In the table, M.I.: 20.028 is the largest, which means that if a connection is established between e24 and e30, then Chi Square will decrease by 20.028. The goal of our model modification is to minimize Chi Square. Therefore eight associations were chosen and added according to the above principle. The model was modified for the second time and run again. The fitting results are shown in Fig. 4.

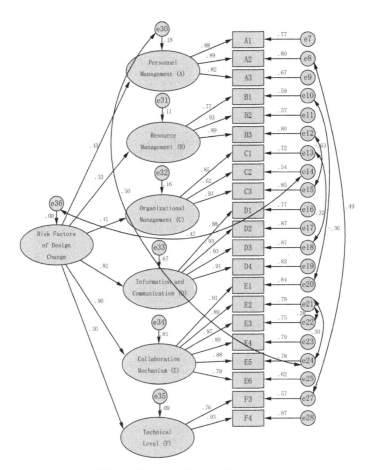

Fig. 4. Second revision of the model.

The coefficient analysis of the model fitting degree index after modification is sorted, as shown in the following Table 8:

According to the coefficient analysis results of the model fitting index, the chi-square degree of freedom ratio (CMIN/DF) is 1.236, which is better than the first modified model. Other indicators have also improved, and the "general" index has been reduced to one. Generally speaking, the model fitting condition is more convincing than before. After modification, all the estimated values of confirmatory factor analysis parameters of the observed variables are within the standard range, and the standard path coefficients of the model are shown in the following table:

Table 8. Model fitting index coefficient analysis after modification.

Fitness test index	Ideal standard	General standard	Model results	Conclusion
CMIN/DF	(1,2)	[2,3)	1.236	Good
RMSEA	<0.08	<0.1	0.048	Good
RMR I RMR	<0.08	<0.1	0.071	Good
GFI	>0.90	>0.8	0.841	General
CFI	>0.90	>0.8	0.978	Good
IFI	>0.90	>0.8	0.979	Good
PNFI	>0.50		0.81	Good

Table 9. Standard path coefficients of the model.

			Estimate	S.E.	C.R.	P	Whether it is established or not
Personnel management	<---	Design change risk	0.43				Establishment
Resource allocation	<---	Design change risk	0.331	0.268	2.41	0.016	Establishment
Organize production	<---	Design change risk	0.406	0.323	2.791	0.005	Establishment
Information and experience	<---	Design change risk	0.82	0.462	3.669	***	Establishment
Collaboration mechanism	<---	Design change risk	0.901	0.55	3.801	***	Establishment
Technical scope	<---	Design change risk	0.302	0.138	2.005	0.045	Establishment

*** means P value is less than 0.001; The significance level was 0.05

As shown in Table 9, the significance level is good, all assumptions are valid, and the overall fitness of the model reaches good after two revisions.

4.4 Analysis of Model Results

The model results show that the research hypothesis H1–H6 are all valid, and the three irrelevant factors F1, F2 and F5 are excluded from the 24 risk factors, and the other factors will induce the risk of design change. The Table 10 below ranks the risk factor levels of design change for each uncertainty: E > D > A > C > B > F.

Table 10. Risk factor levels.

Factors fall into broad categories	Uncertainty	Path coefficient	Description	Factor level
Partnerships	Collaboration mechanism factors (E)	0.901	Closely related and must be adopted	Level I risk factors
	Information and experience (D)	0.82		Level I risk factors
Integrated management capability	Personnel management capacity (A)	0.43	Closely related and need to adopt	Level II risk factors
	Organizational management factors (C)	0.406		Level I risk factors
	Resource allocation status (B)	0.331	General relevance and admissible	Level III risk factors
Technical application level	Technical scope level (F)	0.302		Level III risk factors

5 Proposed Risk-Respond Strategies

Strategies are proposed based on two dimensions: a) characteristics of project, and b) priority of risk factors. There are many possible design change risk events in a project, and the survey also shows that stakeholders are sensitive to the project cost factors, so how to effectively and directionally carry out design change risk control is very important in the case of limited funds and time. If the characteristics of project involve different strategies, then the respond strategies should follow the priorities of different levels, that is Level1—Level2—Level3. High level strategy should be respond priority and then low level strategy could be considered under the time and budget limitation (Table 11).

Table 11. The levels of management strategy.

Levels of strategy	Strategy	Characteristic of project	Risk events	Specific measures
Level 1 strategy	Optimize cooperation mechanism	The partners collaborate at the first time	R7, R10, R11	Optimize cooperation mechanism

(*continued*)

Table 11. (*continued*)

Levels of strategy	Strategy	Characteristic of project	Risk events	Specific measures
				The cooperation plan shall be formulated in a reasonable and strict manner
		Separation work during all the processing such as design, production and assembly	R4, R6, R10, R11	Joint participation
				Follow up actively
		The cooperation mechanism is unreasonable and the terms of the contract are vague	R4–R10	Building trust
				Sign relevant clear and reasonable contract documents
				Establishing a risk-sharing mechanism
	Improve communication efficiency and team experience	Stakeholders have low frequency of communication and traditional inefficiency of information sharing	R2; R9–11	The way information is shared needs to be improved
				Regular and effective communication
				Communicate through joint office or videoconferencing
				Regular exchange of project progress and changes
				Share project experience
				Information sharing has systemic integrity
		Stakeholders' working time is short	R3–6; R8–9	Enhance experience sharing
				Identify change events in similar cases

(*continued*)

Table 11. (*continued*)

Levels of strategy	Strategy	Characteristic of project	Risk events	Specific measures
				Establish a risk pool
Level 2 strategy	Promoting the management mechanism of professional talents	Inappropriate planning and arrangement of managers, and management of multiple projects at the same time, the degree of attention is low	R3–10	Implement a system of rewards and punishments
				Cultivate their enthusiasm for work
				Correct attitude
				Carry out activities or forums to disseminate professionalism
		The talent reserve is insufficient and the speciality is out of tune	R4, R6, R8	Enhance professional level and professional accomplishment
				Performance appraisal, motivation attitude and professional competence
				Conduct professional training and study regularly to strengthen operation skills
				Professional training for potential specialized talents
				Introduce professionals in economic management, wind control and technology
				Cooperate with vocational schools and institutions of higher learning to train and reserve professional talents
	Optimizing the Construction of Organizational Mechanism	Insufficient organizational flexibility, conflict of organizational ideas and non-standard organizational system standards	R3–R11	Develop detailed normative guidelines

(*continued*)

Table 11. (*continued*)

Levels of strategy	Strategy	Characteristic of project	Risk events	Specific measures
				Regular dissemination of organizational culture
				Clear project objectives and organizational strategy
				Establishment of a Design Change Risk Control Expert Committee
Level 3 strategy	Optimize resource allocation	The income distribution mechanism is not scientific and the allocation of resources is unreasonable	R1, R3, R4, R7	Rational and Optimal Allocation of Resources
				Reasonable allocation of time nodes in each phase
				Resource sharing mechanism
	Technical upgrading	Deepening the design level or immature component disassembly technology	R4–R9	Establish a technical expert group to conduct feasibility study and research, and make professional evaluation on the matching degree between the technical scheme and the actual project
				Improve access to the technical market, monitor the whole process of the project, carry out product background tracking, and timely feedback on problems

6 Conclusion

This paper has investigated the project design change risk factors and the control strategy, and discusses the events associated with various common design change risks. Briefly summarized, the design changes arise due to: the addition/reduction of a project, adjustment and change of building function, design scheme changes under detailed design production, assemble, cooperation. In addition, 24 risk factors are selected, which can be summarized into personnel management factors, resource allocation factors, organizational management factors, information and experience factors among partners, cooperation mechanism factors and technical level factors. And the relationship between factors and events is also sorted out. The structural equation model was used to study the risk control strategy of design change in prefabricated projects, and a theoretical model was established. Data collected using questionnaires were used to fit the model and verify all assumptions. In the model, 21 uncertain factors were identified, 3 unrelated factors were excluded, and the risk factors were ranked as follows: collaboration mechanism factors > experience and communication > personnel management ability > organization and management factors > the condition of resource configuration > technology application, and develop risk management strategies for design change according to the different risk characteristics, as well as the level of risk factors.

This paper focuses on the specific problem of design change risk, and uses an empirical study to study the impact of design change risk factors, identify and prioritize them for different design change risk events. The proposed specific management strategies will help to reduce the delay of project progress, reduce costs and maintain normal working procedures. In the following research, we can apply the strategy to actual projects and collect feedback to further validate and modify the model.

Acknowledgment. This work is supported by the Chinese Ministry of Education of Humanities and Social Science Project under Grant 17YJC630021.

References

Ahmadabadi, A.A., Heravi, G.: Risk assessment framework of PPP-megaprojects focusing on risk interaction and project success. Trans. Res. Part A: Policy Pract. **124**, 169–188 (2019)

Du, J., Jing, H., Castro-Lacouture, D., et al.: Multi-agent simulation for managing design changes in prefabricated construction projects. Eng. Constr. Architectural Manage. **27**(1), 270–295 (2019)

Layth Kraidi, R.S., Wilfred Matipa, F.B.: Analyzing the critical risk factors associated with oil and gas pipeline projects in Iraq. Int. J. Crit. Infrastruct. Prot. **24**, 14–22 (2019)

Retik, A., Warszawski, A.: Automated design of prefabricated building. Build. Environ. **29**(4), 421–436 (1994)

Qi, B.K., Li, C.F.: Whole life cycle management of prefabricated construction research based on BIM technology. Appl. Mech. Mater. **536–537**, 1705–1708 (2014)

Sheng-Hua, Z.: Research on the fundamental framework of alliance capability and its mechanism of promoting alliance performance. In: 2006 International Conference on Management Science and Engineering, Lille, pp. 963–969 (2006)

Tong, Q., Liu, T., Tong, H., Ou, Z.: Evaluation of historic buildings based on structural equation models. In: Shi, Y., van Albada, G.D., Dongarra, J., Sloot, Peter M.A. (eds.) ICCS 2007. LNCS, vol. 4489, pp. 162–165. Springer, Heidelberg (2007). https://doi.org/10.1007/978-3-540-72588-6_28

Wang, T., Gao, S., Li, X., et al.: A meta-network-based risk evaluation and control method for industrialized building construction projects. J. Clean. Prod. **205**, 552–564 (2018)

Yue, Z., Han, P., Ling, Z.: Risk analysis of coal mine construction project based on structural equation model. In: 2010 3rd International Conference on Information Management, Innovation Management and Industrial Engineering, Kunming. pp. 23–26 (2010)

Yang, J., Wang, J.: Research on construction project integrated management model. Constr. Econ. **03**, 67–69 (2008)

AHP-FCE Evaluation of Cross-Border e-Commerce Supply Chain Performance for Xi'an International Inland Port

Guo-Ling Jia[✉]

Xi'an International University, No.18 YuDou Road, Yanta District, Xi'an, China

Abstract. Cross-border E-commerce has gained great development with the support of Xi'an international inland port. It is significant to evaluate the supply chain performance to promote the better development. The combination of Analytical Hierarchy Process and Fuzzy Comprehensive Evaluation is introduced to carry out the evaluation. Results show that the satisfaction is relatively high and there still is some room for the Xi'an international inland port to improve the supply chain performance, mainly concentrating on supply chain responsiveness and logistics service.

Keywords: Analytical hierarchy process · Fuzzy comprehensive evaluation · Xi'an international inland port · Cross-border e-commerce · Supply chain

1 Introduction

In the background of global trade, cross-border e-commerce (CBEC) has gained increasingly development and become a trending issue in China (Center 2018). Most of the traditional e-commerce and retail companies have begun the CBEC business to meet the needs of global customers. The trade volume of China's CBEC expanded greatly recently as showed in Figs. 1 and 2. Therefore, the competition between these companies is getting more and more intensive.

Xi'an International Inland Port (XAIIP) is an important export-oriented economic platform for Xi'an city and Shaanxi province. It has prominent advantages to develop CBEC industry. Early in march, 2014, the National General Administration of Customs officially approved Xi'an as a pilot city for CBEC services. And the Xi'an Municipal government relied on the XAIIP to carry out CBEC pilot projects. In October 2014, in order to provide better service for CBEC enterprises, XAIIP developed Xi'an CBEC platform and officially launched. The platform integrates logistics, warehousing, distribution, customs declaration, inspection and quarantine. Third-party payment and other functions provide one-stop integrated services for CBEC companies entering the region. From 2015 to 2018, the import and export volume has maintained a growth rate of more than 30% in successive years, and there are more than 200 registered enterprises. A number of well-known CBEC companies such as Dunhuang Network and Alibaba.com have been cross-border purchases. Local high-quality enterprises such as Silk Road City have

© Springer Nature Switzerland AG 2020
K. R. Lang et al. (Eds.): WeB 2019, LNBIP 403, pp. 215–226, 2020.
https://doi.org/10.1007/978-3-030-67781-7_19

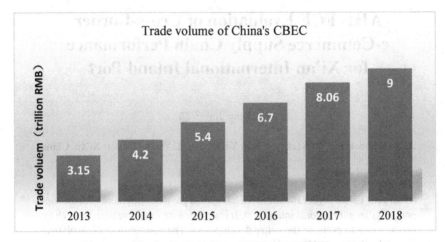

Fig. 1. Trade volume of China's CBEC

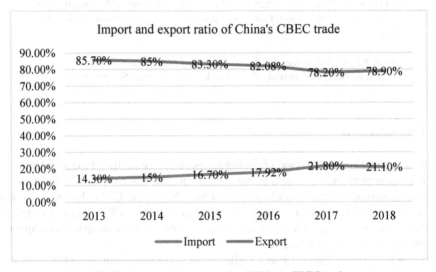

Fig. 2. Import and export ratio of China's CBEC trade

successively started business. The port has completed more than 10 million cross-border import and export orders, of which more than 90% of export orders accounted for more than 200 million US dollars, effectively driving the development of CBEC industry in Shaanxi and Xi'an, and effectively supporting the development of cross-border industry in the western region.

As few evaluation work has been done for the inland port in the context of CBEC, the main purpose of this paper is to evaluate the CBEC supply chain performance of XAIIP, and provide practical implications for the platform owner to deal with the evaluation result.

The remainder of the paper is organized as follows. Section 2 describes the related studies; Sect. 3 introduces evaluation method and evaluation indicator frame; Sect. 4 present the findings of the evaluation; Sect. 5 offers the discussion and implication; and finally Sect. 6 concludes the study.

2 Literature Review

Studies on CBEC supply chain can be classified as the following 4 types. Firstly, researchers began to realize the new challenges for CBEC supply chain. Cho, Boo-yun identify the influence of the manufacturers' efforts for e-business channel enhancements along with traditional collaboration with local retailers (Cho and 이성희 2014). Xu, Yaqing et al. discussed a cross-border supply chain system that has an import e-commerce company and overseas supplier (Xu, Gui et al. 2019). Su, Weihua et al. explored the impacts of supervision of payments and transactions on CBEC supply chain, quantitatively analyzed the internal structure and dynamic layout characteristics of sustainable CBEC policy documents (Su, Wang et al. 2019). Kawa Arkadiusz concentrated on the factors of delivery of products on cross-border flow and develop a model facilitating cooperation between online shops dealing with cross-border trade (Kawa 2017). Secondly, according to the problems occurred with the requirement of CBEC, some researchers put forward supply chain optimization program to deal with the changing environment. Razmerita Liana et al. present an advanced e-custom infrastructure designed to streamline trade procedures, prevent potential security threats and reduce tax related fraud in cross-border trade (Razmerita and Bjorn-Andersen 2008). Chen Jengchung Victor et al. proposed two additions to the existing IS success model to influence usefulness and satisfaction, and ultimately customer loyalty (Chen, Chen et al. 2013). Ding Feng et al. build a game model to analyze how consumers preferences affect the competitive strategy making in cross-border dual-channel supply chain (Ding, Chen et al. 2019). The optimal pricing and service level under centralized and decentralized decision are formulated. Wang Ying et al. investigate Zongteng companies to assess how supply chain innovation activities can become core to a firm's business model innovation (Wang, Jia et al. 2018). Zhang Hao-Zhan et al. analyzed service risk of cross-border logistics based on Supply Chain Continuous Operation Management and developed the strategies that aim at the two factors above and propose some ways to reduce logistics service risk (Zhang, Hsieh et al. 2017). Song,Bo proposed a knowledge-based CBEC risk assessment system and aims at improving system functions including commodity classification and knowledge acquisition (Song, He et al. 2017). Zhao Lianming established the optimization model of low-carbon closed-loop supply chain network with multiple objectives. A kind of improved genetic algorithm combined with epsilon constraint method is put forward to get the Pareto optimal solution (Zhao 2016). Thirdly, alterative CBEC supply chain models were proposed to be consistent with the different product. Liu Yinghan discussed the logistics pattern and the logistics development of cross-border e-commerce (Liu 2017). Feng Lipeng et al. proposed L-COPRAS method to deal with uncertain or linguistic expression on strategic cost measures with varied weights to different alternatives (Feng, Ma et al. 2017). Lastly, evaluation of CBEC supply chain. Ma Jun et al. developed a Complex proportion assessment method and used complex fuzzy set to tackle uncertainty and temporal features in given evaluation context (Jun, Lipeng et al. 2017).

To sum up, the evaluation of CBEC supply chain has not been covered largely. To the knowledge of the author, few of the studies focused on Xi'an international Inland Port CBEC supply chain.

The focus of competition is actually about the supply chain service quality. The suitable supply chain has the ability to comply with the requirements of companies and customers and integrate resources. With the purpose of optimizing the cross-border e-commerce supply chain service ability, it is important to evaluate its performance at first. The evaluation of CBEC supply chain is significant to Xi'an inland port and help to improve its service level.

3 Method

In terms of evaluation, there are many popular methods that can be used such as Delphi method, Analytic Hierarchy Process (AHP), Factor analysis, Cluster analysis, Artificial Neural Network (ANN), Grey Comprehensive Evaluation (GCE) and Fuzzy Comprehensive Evaluation method (FCE). Main differences are showed in Table 1. In general, some methods have obvious subjectivity, while some ask for high level of samples. In contrast, FCE can avoid the shortcomings of "unique solution" for mathematical methods. Meanwhile, cross-border e-commerce supply chain is such a complex system

Table 1. Comparison of common evaluation methods

Evaluation method	Strength	Weakness	Suitable case
Delphi	Easy to use, professional experience are combined	Obvious subjectivity; Hard to converge	Strategic level decision; systems hard to quantify
AHP	Easy to use	Strong subjectivity	Cannot provide new project; less quantitative data and more qualitative
Factor analysis	Objective and reasonable	Data volume and composition are strictly required	Classification
Cluster analysis	Deal with the relationship between different object	Large quantity of data are required	Market segmentation; consumer behavior division; sampling plan design
ANN	Deal with large, complex systems that are nonlinear, non-local and non-convex	Low accuracy and large number of samples are required	Stock price assessment; Comprehensive evaluation of urban development

(*continued*)

Table 1. (*continued*)

Evaluation method	Strength	Weakness	Suitable case
GCE	Easy calculation	Limited to compare pros and cons of the object	Problems with less data and uncertain information
FCE	Make a scientific and reasonable quantitative evaluation with ambiguous information	Cannot solve the duplication problem of relevant factors	Solving various non-deterministic problems

that the recognition to it is still uncertain. And some of the indicators can not be exactly quantified. FCE is favorable for the fuzzy information and can give more feasible results.

FCE is one of the most widely used methods in fuzzy mathematics such as education evaluation, economic evaluation and management evaluation. AHP is a popular method used for indicator weight allocation. This paper combined the two methods to evaluate the CBEC supply chain performance for XAIIP.

The AHP-FCE mainly includes two parts. In the first part, AHP is used to establish the evaluation indicator frame. A pairwise comparison judgment matrix is constructed, and the weight vector of each indicator is calculated. As for the second part, FCE is used to carry out the evaluation based on the first part. The combination of these two methods help to improve the reliability and validity of the evaluation results.

3.1 Constructing Bipartite Judgment Matrix

In the first part, constructing a bipartite judgment matrix is the key process. According to AHP, the importance degree of each factor in each level is determined by comparison. The scale and its meaning is showed in Table 2.

Table 2. Importance comparison score and its explanation

Scale	Explanation
1	Equally influential
3	Slightly influential
5	Strongly influential
7	Strongly influential
9	Absolutely influential
2	Sporadic values between two close scales
4	
6	
8	

3.2 Consistency Check

After getting the weight value in the hierarchical order, we need to check the consistency. This is finished by calculating the Consistency Index at first.

$$CI = \frac{\lambda_{max} - n}{n - 1} \tag{1}$$

Then, look up the corresponding average random Consistency Correction Factor RI (Random Index). Lastly, calculate the Consistency Ratio (CR).

$$CR = \frac{CI}{RI} \tag{2}$$

If $CR < 0.10$, the consistency can meet the requirements.

3.3 FCE Evaluation Steps

Step 1: Construct the indicator set of the evaluation object. Given the number of indicator is n, so

$$A = (A_1, A_2, A_3, \ldots, A_n) \tag{3}$$

Step 2: Construct the comment set. Suppose the number of comment level is m, then the comment set is showed as Eq. (4).

$$V = (v_1, v_2, \ldots, v_n) \tag{4}$$

The comment is excellent, good, mean, bad, very bad respectively.

Step 3: Construct the fuzzy relational matrix. The fuzzy subset is established in the previous step and it is necessary to quantify the evaluation items. So the fuzzy relation matrix is obtained by calculating the degree of membership of each indicator to fuzzy subset.

$$R = \begin{Bmatrix} r_{11} \ldots r_{1m} \\ \vdots \ddots \vdots \\ r_{n1} \ldots r_{nm} \end{Bmatrix} \tag{5}$$

Where r_{ij} is the degree of membership of indicator A_i to comment level v_j.

Step 4: Get the weight vector of indicator.

This is calculated by the means of AHP.

Step 5: Calculate the fuzzy composite value.

The fuzzy comprehensive evaluation results are obtained by Eq. (6).

$$B = W \cdot R = (w_1, w_2, \ldots w_m) \begin{Bmatrix} r_{11} \ldots r_{1m} \\ \vdots \ddots \vdots \\ r_{n1} \ldots r_{nm} \end{Bmatrix} = (b_1, b_2, \ldots b_n) \tag{6}$$

3.4 Evaluation Indicator Frame

As product flow and storage play a significant role in the CBEC, this paper takes XAIIP logistics platform as the center of CBEC supply chain to set up evaluation indicator frame. The main purpose of CBEC supply chain is to meet the consumers' requirement at the right time and right cost. The smooth process of supply chain is key for the order processing. Supply chain cost can reflect the efficiency and profit level. Responsiveness of CBEC supply chain is critical for the customers of e-commerce compared to the traditional business. Logistics platform service level affects the delivery time and cost for CBEC companies. So, five primary indicators are set up and their secondary indicators are extended according to the particular operation as showed in Table 3.

Table 3. CBEC supply chain evaluation indicators

Objective	Primary indicator	Secondary indicator
CBEC Supply Chain evaluation indicators A	Customer satisfactory A_1	Customer satisfaction A_{11}
		Customer complaints rate A_{12}
		Return rate A_{13}
		Product quality A_{14}
	Supply chain process A_2	Information share A_{21}
		Cohesion A_{22}
		Traceability A_{23}
	Supply chain cost A_3	Inventory turnover A_{31}
		Profit rate A_{32}
		Resource consumption A_{33}
		Environment pollution A_{34}
	Supply chain responsiveness A_4	Responsive time A_{41}
		Order fulfillment rate A_{42}
		Delivery accuracy A_{43}
		Information cover range A_{44}
	Logistics service level A_5	Bonded warehouse A_{51}
		Customs clear A_{52}
		Inspection and quarantine A_{53}
		Third-party payment A_{54}
		Delivery A_{55}

4 Findings

The survey on the supply chain performance of CBEC in XAIIP is conducted among users on website. The respondents are limited to employees or owners of CBEC enterprises.

The types of enterprises who took part in the survey are presented as Table 4. Overall, 150 respondents are invited to do the survey and 127 questionnaires are selected after invalid ones are removed. SPSS software are used to test the reliability and validity of the collected samples. The Cronbach's Alpha of reliability testing is 0.896, showing the questionnaire with internal consistency. The popular KMO test value is 0.701, so the sample data is effective. The evaluation results are showed in Table 5.

Table 4. The type of interviewed CBEC enterprises

Type	Quantities	Ratio
Management	29	22.83%
Operation	43	33.86%
Technology	25	19.69%
Logistics	30	23.62%
Total	127	100%

Then, the fuzzy relation matrix is obtained as followings.

$$R_1 = \begin{Bmatrix} 0.3371 & 0.4270 & 0.2360 & 0.0000 & 0.0000 \\ 0.3034 & 0.6966 & 0.0000 & 0.0000 & 0.0000 \\ 0.2135 & 0.1910 & 0.5955 & 0.0000 & 0.0000 \\ 0.5618 & 0.3258 & 0.1124 & 0.0000 & 0.0000 \end{Bmatrix}$$

$$R_2 = \begin{Bmatrix} 0.5056 & 0.3371 & 0.2697 & 0.0000 & 0.0000 \\ 0.4494 & 0.3371 & 0.3258 & 0.0000 & 0.0000 \\ 0.3596 & 0.3933 & 0.2472 & 0.0000 & 0.0000 \end{Bmatrix}$$

$$R_3 = \begin{Bmatrix} 0.2809 & 0.5056 & 0.2135 & 0.0000 & 0.0000 \\ 0.3371 & 0.3371 & 0.3258 & 0.0000 & 0.0000 \\ 0.2022 & 0.5393 & 0.2584 & 0.0000 & 0.0000 \\ 0.1348 & 0.2247 & 0.6405 & 0.0000 & 0.0000 \end{Bmatrix}$$

$$R_4 = \begin{Bmatrix} 0.2360 & 0.2584 & 0.3034 & 0.2022 & 0.0000 \\ 0.2697 & 0.3371 & 0.3371 & 0.0225 & 0.0000 \\ 0.0674 & 0.6742 & 0.2560 & 0.0225 & 0.0000 \\ 0.2247 & 0.5618 & 0.1348 & 0.7866 & 0.0000 \end{Bmatrix}$$

$$R_5 = \begin{Bmatrix} 0.3371 & 0.5618 & 0.1011 & 0.0000 & 0.0000 \\ 0.1685 & 0.4494 & 0.3371 & 0.0449 & 0.0000 \\ 0.2472 & 0.5281 & 0.2247 & 0.0000 & 0.0000 \\ 0.2584 & 0.5056 & 0.2022 & 0.0337 & 0.0000 \\ 0.1910 & 0.4607 & 0.3483 & 0.0000 & 0.0000 \end{Bmatrix}$$

Table 5. Scoring ratio of individual indicator

Primary indicator	Secondary indicator	Excellent	Good	Mean	Bad	Very bad
A_1	A_{11}	0.3371	0.4270	0.2360	0.0000	0.0000
	A_{12}	0.3034	0.6966	0.0000	0.0000	0.0000
	A_{13}	0.2135	0.1910	0.5955	0.0000	0.0000
	A_{14}	0.5618	0.3258	0.1124	0.0000	0.0000
A_2	A_{21}	0.5056	0.3371	0.2697	0.0000	0.0000
	A_{22}	0.4494	0.3371	0.3258	0.0000	0.0000
	A_{23}	0.3596	0.3933	0.2472	0.0000	0.0000
A_3	A_{31}	0.2809	0.5056	0.2135	0.0000	0.0000
	A_{32}	0.3371	0.3371	0.3258	0.0000	0.0000
	A_{33}	0.2022	0.5393	0.2584	0.0000	0.0000
	A_{34}	0.1348	0.2247	0.6405	0.0000	0.0000
A_4	A_{41}	0.2360	0.2584	0.3034	0.2022	0.0000
	A_{42}	0.2697	0.3371	0.3371	0.0225	0.0000
	A_{43}	0.0674	0.6742	0.2560	0.0225	0.0000
	A_{44}	0.2247	0.5618	0.1348	0.7866	0.0000
A_5	A_{51}	0.3371	0.5618	0.1011	0.0000	0.0000
	A_{52}	0.1685	0.4494	0.3371	0.0449	0.0000
	A_{53}	0.2472	0.5281	0.2247	0.0000	0.0000
	A_{54}	0.2584	0.5056	0.2022	0.0337	0.0000
	A_{55}	0.1910	0.4607	0.3483	0.0000	0.0000

$$R = \left\{ \begin{array}{c} R_1 \\ R_2 \\ R_3 \\ R_4 \\ R_5 \end{array} \right\}$$

The weight of primary indicators and secondary indicators calculated by the above mentioned AHP are w, w_1, w_2, w_3, w_4, w_5.

$$w = (0.2313, 0.1716, 0.2910, 0.1567, 0.2239)$$
$$w_1 = (0.3057, 0.2576, 0.1441, 0.2926)$$
$$w_2 = (0.2941, 0.0118, 0.3176)$$
$$w_3 = (0.2564, 0.2949, 0.1795, 0.2051)$$
$$w_4 = (0.2961, 0.1974, 0.2632, 0.2237)$$
$$w_5 = (0.2564, 0.2009, 0.2222, 0.1709, 0.1496)$$

Next, the fuzzy composite value can be obtained by the following calculation.

$$B_1 = w_1 \cdot R_1 = (0.3057, 0.2576, 0.2360, 0, 0)$$
$$B_2 = w_2 \cdot R_2 = (0.3176, 0, 3176, 0.2687, 0, 0)$$
$$B_3 = w_3 \cdot R_3 = (0.2949, 0.2949, 0.2949, 0, 0)$$
$$B_4 = w_4 \cdot R_4 = (0.2360, 0.2584, 0.2961, 0.2237, 0)$$
$$B_5 = w_5 \cdot R_5 = (0.2564, 0.2564, 0.2009, 0.0449, 0)$$

Normalized the B_1, B_2, B_3, B_4 and B_5, then

$$B_1' = (0.3825, 0.3222, 0.2953, 0, 0)$$
$$B_2' = (0.3514, 0.3514, 0.2862, 0, 0)$$
$$B_3' = (0.3333, 0.3333, 0.3333, 0, 0)$$
$$B_4' = (0.2327, 0.2548, 0.2920, 0.2206, 0)$$
$$B_5' = (0.3380, 0.3380, 0.1981, 0.0519, 0)$$

$$B = W \cdot R = (w_1, w_2, \ldots w_5) \left\{ \begin{pmatrix} r_{11} & \cdots & r_{15} \\ \vdots & \ddots & \vdots \\ r_{51} & \cdots & r_{55} \end{pmatrix} \right\} = (b_1, b_2, \ldots, b_n) = (0.2137, 0.5219, 0.2026, 0.0168, 0)$$

The evaluation of each indicator is showed in Table 6.

Table 6. The FCE result of each indicator

Indicator	Excellent	Good	Average	Poor	Very Poor
Customer satisfactory A_1	38.25%	32.22%	29.53%	0	0
Supply chain process A_2	35.14%	35.14%	28.62%	0	0
Supply chain cost A_3	33.33%	33.33%	33.33%	0	0
Supply chain responsiveness A_4	23.27%	25.48%	29.20%	22.06%	0
Logistics service level A_5	33.80%	33.80%	19.81%	5.19%	0

5 Discussion and Implication

The above study results have some practical implication for the XAIIP supply chain platform organizations. According to the overall performance evaluation, the fuzzy membership of the supply chain performance include excellent, good, average, poor, very poor. The correspondent ratio is 21.37%, 52.19%, 29.26%,1.68% and 0 respectively. It suggests that around two-thirds of the CBEC enterprises make a positive evaluation for the service of XAIIP platform, with 21.37% for excellent and 52.19% for good.

As for each specific indicator A_1 to A_5, we can see from Table 5 that the ratio of former three items are basically even, and there are no poor performance. This means

XAIIP have relative advantages in these three indicator, but not so distinguished. There still is potential for the customer satisfactory, supply chain process and supply chain cost to be promoted. As for the forth indicator supply chain responsiveness A_4, there is some difference between it and the former there ones. Obviously, 22.06% of the users show unsatisfactory assess. As the process of the CBEC is most likely to be prolonged due to the lack of advanced information system and optimized process, the responsiveness is affected accordingly. So, the XAIIP should focus more on the responsiveness factors, such as accelerating the customs clearance and investing on information system. Similarly, 5.19% of the users are not satisfied with the logistics service level A_5. There is certain relationship between A_4 and A_5. Logistic service level has some impactions on supply chain responsiveness. If the XAIIP improve logistics service level, it will facilitate the supply chain responsiveness. It can be concluded that logistics provider and service quality is a primary issue to be greatly developed. The provider selection and facility improvement can contribute to the optimization of A_5.

6   Remarks and Conclusion

In conclusion, the competition advantage of CBEC rely greatly on its supply chain service level. In order to optimize the performance of CBEC supply chain, the evaluation indicator frame for XAIIP CBEC according to AHP is set up and their weight is calculated, this is followed by the fuzzy comprehensive evaluation method. The evaluation results show that more than half of the evaluators give the high satisfaction for the service of XAIIP. However, there is room for it to promote service level. By reasonable resource distribution, more needs of warehousing, exhibition, transaction can be met. Domestic and foreign well-known supply chain integrated service providers can gather to carry out cross-border trade as well. In future, more studies can be carried out on comparative research among several representative inland port, the evaluation method can be compared and combined to explore inspiring proposals as well.

Acknowledgements. This work is financially supported by Shaanxi Science and Technology funding (grant number 2015SF296) and fund for Innovative Research Team Program of Xi'an International University (grant number XAIU-KT201802-2)

References

Center, E.-c. r.: China Cross-border E-commerce Market Data Report (2018)
Chen, J.V., Chen, Y., Capistrano, E.P.S.: Process quality and collaboration quality on B2B e-commerce. Ind. Manage Data Syst. **113**(6), 908–926 (2013)
Cho, B.-Y., 이성희, : The effects of e-business channel enhancement for foreign market entry through comparison between two cases: independent and vertically integrated off-line channels. J. Internet Electron. Commer. Res. **14**(4), 355–375 (2014)
Ding, F., Chen, J., Chen, C., Huo, J.: Study on cross-border e-commerce competitive differentiation strategy. Oper. Res. Manage. Sci. **28**(6), 33–40 (2019)
Feng, L., Ma, J., Wang, Y., Yang, J.: Supply chain downstream strategic cost evaluation using L-COPRAS method in cross-border E-commerce. Int. J. Comput. Intell. Syst. **10**(1), 815–823 (2017)

Jun, M., Lipeng, F., Jie, Y.: Using complex fuzzy sets for strategic cost evaluation in supply chain downstream (2017)

Kawa, A.: Supply chains of cross-border e-commerce. In: Król, D., Nguyen, N.T., Shirai, K. (eds.) ACIIDS 2017. SCI, vol. 710, pp. 173–183. Springer, Cham (2017). https://doi.org/10.1007/978-3-319-56660-3_16

Liu, Y.: Study on logistics pattern of cross-border E-commerce. In: Jing, W., Guiran, C., Huiyu, Z. (eds.) Proceedings of the 2016 7th International Conference on Education, Management, Computer and Medicine, vol. 59, pp. 884–887 (2017)

Razmerita, L., Bjorn-Andersen, N.: A service-oriented infrastructure for advanced cross-border trade (2008)

Song, B., He, J., Yan, W., Hu, Q., Zhang, T.: Key technologies for knowledge-based cross-border e-commerce risk assessment - accurate commodity classification and efficient knowledge acquisition. In: Chen, C.H., Trappey, A.C., Peruzzini, M., Stjepandic, J., Wognum, N. (eds.) Transdisciplinary Engineering: A Paradigm Shift, vol. 5, pp. 146–153 (2017)

Su, W., Wang, Y., Qian, L., Zeng, S., Balezentis, T., Streimikiene, D.: Creating a sustainable policy framework for cross-border E-commerce in China. Sustainability 11(4), 943 (2019)

Wang, Y., Jia, F., Schoenherr, T., Gong, Y.: Supply chain-based business model innovation: the case of a cross-border E-commerce company. Sustainability 10(12), 4362 (2018)

Xu, Y., Gui, H., Zhang, J., Wei, Y.: Supply chain analysis of cross border importing e-commerce considering with bonded warehouse and direct mailing. Sustainability 11(7), 1909 (2019)

Zhang, H.-Z., Hsieh, C.-M., Luo, Y.-L., Chiu, M.-C.: An investigation of cross-border E-commerce logistics and develop strategies through SCCOM framework and logistic service risk analysis. In: Chen, C.H., Trappey, A.C., Peruzzini, M., Stjepandic, J., Wognum, N. (eds.): Transdisciplinary Engineering: A Paradigm Shift, vol. 5, pp. 102–113 (2017)

Zhao, L.: Evaluation system construction for logistics supply chain of cross-border E-commerce. In: Fang, Y., Xin, Y. (eds.) Proceedings of the 2016 4th International Conference on Machinery, Materials and Information Technology Applications, vol. 71, pp. 1563–1569 (2016)

Knowledge Domain and Emerging Trends in Cross-Border E-commerce Coordination Mechanism Based on CiteSpace Analysis

Shan Du[✉] and Hua Li

School of Economics and Management, Xidian University, Xi'an, China
dushan_19860617@126.com, lihua@xidian.edu.cn

Abstract. With the advance of One Belt and One Road initiative (BRI), cross-border e-commerce has experienced rapid growth and needs urgent attention from researchers. We analyze the literature from the SCI-EXPANDED, SSCI, CPCI-S, CPCI-SSH database on cross-border e-commerce coordination mechanism to find the knowledge domain and emerging trends in the new background. In order to help researchers and practitioners grasp the research frontier in this field quickly. We develop a framework of cross-border e-commerce coordination mechanism by analyzing the most influential authors, institutions, countries, keywords and references. We apply knowledge mapping cluster view to our study. Frequency statistics, clustering coefficient as well as centrality calculation are employed to analyze by CiteSpace. We use the strength of citation bursts to analyze keywords and present the major clusters to reveal their associated intellectual bases. In this study, we explore the knowledge structure, development and the future trend of cross-border e-commerce coordination mechanism for researchers. We identify the main technology and models to improve cross-border e-commerce participant satisfaction.

Keywords: Cross-border e-commerce · CiteSpace · Coordination mechanism · Emerging trend

1 Introduction

In recent years, cross-border E-commerce has become a hotspot. Development of cross-border E-commerce makes traditional export trade structure change, more and more foreign trade enterprises acquire business opportunities through continuous online transactions. Chinese cross-border e-commerce transactions will continue to develop with high-speed. Cross-border e-commerce transactions reached 6.4 trillion, and it accounts for the proportion of China's import and export trade will be increasing to reach 18.5% by 2020. 1% increase in the use of efficient and flexible cross-border payment systems could increase cross-border e-commerce by as much as 7%. E-commerce cross-border trade is unfolding and there is a huge market potential (Lai and Wang 2014).

© Springer Nature Switzerland AG 2020
K. R. Lang et al. (Eds.): WeB 2019, LNBIP 403, pp. 227–235, 2020.
https://doi.org/10.1007/978-3-030-67781-7_20

As the competition in the cross-border E-commerce industry is increasingly fierce, the cross-border E-commerce industry has been subdivided into multiple operation modes gradually. In order to help the participants of cross-border E-commerce industry clarify the development status of Chinese cross-border E-commerce, and learn to deal with various specific issues in practice. After comparative analysis on typical cross-border E-commerce coordination mechanism literature, it is concluded that cross-border E-commerce has more potential to improve in aspects of logistics, storage, payment, policy and talent. Only in this way can cross-border E-commerce enterprises seize opportunity to develop rapidly and make the cross-border E-commerce become the driver of cross-border trade growth (Ai et al. 2016). BRI is likely to increase the complexity of cross-border trade. Firms will respond to this change in different ways, and how they respond may ultimately define who the winners and losers are in the BRI era.

The article is structured as follows. The next section presents the related work that we research, including the concepts and common mechanisms. Section 3 explains the approach for visual analysis. The results that we obtained from the CiteSpace are explained in Sect. 4. Finally, Sect. 5 summarizes and presents some policy-related conclusions.

2   Related Work

Bibliometrics is the most common statistical method to analyze literature. The main categories of bibliometrics are the authors, keywords, references, journals and the trends. Visualized analysis of bibliometrics can be proceeded under the computer technology. As an important branch, visual analysis software has achieved significant progress. Mapping knowledge as a useful method has been widely applied in revealing the trend of some fields. Visualized analysis based on co-citation can provide more usefull information. It could help researchers find the recent trend in a special field and forecast the future direction. We present the development of cross-border e-commerce coordination mechanism by using CiteSpace which supports by Java application. CiteSpace was identified as a bibliometric analysis tool, which could generate visual maps in order to discover trends in a field. It is designed for finding turning point in the evolution of a pecific domain. It provides a useful literature analysis tool to identify fast-growing areas, find collaborative relationships between institution and country, apply a network into clusters, labell clusters from citing articles. In building the network view, three types of views—cluster view, timeline view and timezone view can be applied to analyze different results including knowledge structure and evolution of trends, respectively. The preferred data source for CiteSpace is the Web of Science. CiteSpace includes nodes and links. Different nodes represent different elements in a visualization knowledge map such as country, reference, institution and author. Links between nodes should be considered as strength of relationships. The size of nodes indicates the aggregate co-occurrence frequency of an element, whereas the color of the ring in a map show the co-occurrence time slices of this item. The nodes with a purple color represents high centrality which usually perceived as critica turning points in a field. Centrality which measures the ability of a node connecting with other nodes. In other words, a node with a high centrality (range from 0 to 1) illustrates that the node acts as a key point linking and the primary topics in a network. We can solve the problem by using citespace, for example, which articles are important turning point in cross-border E-commerce research field. Which subject are mainstream. How has the research front evolve (Synnestvedt et al. 2005).

There are several motives that have been put forward to explain the inception of the BRI, including: allowing for the internationalisation of the Chinese currency and re-balancing of foreign currency reserves; gaining greater and more reliable resources; making better use of excess production capacities in China; and allowing for the development of China's Western provinces. All of the above motives are likely to have a strong impact on the cross-border E-commerce. In fact, gaining access to resources, reducing excess production capacities, and location decisions for logistics center are all directly related to cross-border E-commerce. Thus, the potential impact of the BRI on cross-border E-commerce coordination mechanism theory and practice has been largely overlooked. Further, his is a topic that warrants greater attention from s E-commerce researchers and practitioners both within and outside of China. In response, we use a systematic review of the broader literature on the cross-border E-commerce coordination mechanism. Using these insights as a backdrop, we then seek to explore how the cross-border E-commerce coordination mechanism is changing in the context of BRI.

3 Approach

This paper use CiteSpace and Java to undertake a visual analysis of cross-border E-commerce. We collect data from 1985 to 2019 based on the Web of Science database. Because a lot of high-quality papers and influential articles are collected in the database and it is the preferred database website for CiteSpace. In order to collect the full records of samples and eliminate useless information. First, we chose Web of Science Core Collection. Secondly, selecting index of Science Citation Index Expanded (SCI-EXPANDED), Social Sciences Citation Index (SSCI), Conference Proceedings Citation Index-Science (CPCI-S) and Conference Proceedings Citation Index-Social Science and Humanities (CPCI-SSH) as target set. We search the term of "cross-border e-commerce" or "coordination mechanism" then 1990 articles relate to cross-border e-commerce and coordination mechanism was found. We put the articles into citespace and discover the important information of cross-border E-commerce (Table 1).

Table 1. Summary of searching details.

Source website	Web of science
Database	SCI-EXPANDED; SSCI; CPCI-S; CPCI-SSH
Years	1985–2019
Searching term	"Cross-border e-commerce" or "coordination mechanism"
Sample size	1990

4   Analysis and Results

4.1   Mapping and Analysis on Author

We select the top 10 most productive authors from the output of CiteSpace. Then we serach their articles in the database to find their main focuses. Most of them pay attention to the field of consumer behavior and cross-border E-commerce platform, such as consumer behavior preference, import and export trade, buyer system and industrial cluster. We listed the top 10 most productive authors and their main focuses in Table 2. The map of authors network is presented in Fig. 1, in the map, blue is representative of earlier articles, green is for middle years whereas yellow is viewed as recent articles. Since cross-border E-commerce appears in China relatively late, the articles are fewer than e-commerce. Meanwhile, we have selected the articles from several database, some publications about cross-border E-commerce beyond our database are not be mentioned.

Table 2. Top 10 most productive authors.

Author	Publications	Main focuses
YAO L	7	Cross-border E-commerce platform, consumer behavior preference, import and export trade, buyer, industrial Cluster, consumer research
WANG Y	4	Cultural symbols, logistics Models
YAN W	3	Risk analysis, risk Assessment
WU YL	3	Users Behavior, cluster Analysis, country-based Difference
WANG L	3	Cross-border logistics, transaction costs, supply chains
ZHAO H	2	B2C Cross-border E-commerce, logistics
YANGZH	2	Cultivation mechanism, cross-border E-commerce supervision
YANG J	2	Institutional innovation, business-mode innovation
XUE WX	2	Ecological model, cross-border e-commerce ecosystem
WANG HP	2	2-SIDED MARKETS, cross-border e-commerce comprehensive pilot areas

4.2   Mapping and Analysis on Countries and Institutions

As shown in Fig. 2, the mainland China with the highest degree of centrality published the most relevant articles on cross-border e-commerce. Because of the financial crisis, the export volume of foreign trade decreased significantly, while the cross-border e-commerce industry characterized by "small batch and high frequency" achieved rapid development. According to the data of China e-commerce research center, the cross-border e-commerce transaction volume in China reached 7.3 trillion in 2018. China was also the first country to put forward the "One Belt And One Road" initiative.

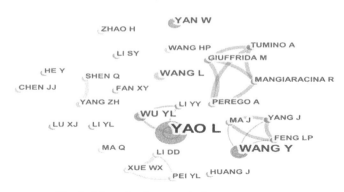

Fig. 1. Map of authors co-citation.

The rapid development of cross-border e-commerce attracts more and more scholars' attention. Some European countries have also made great contributions to cross-border e-commerce, such as Australia, the United States and the Italy.

Fig. 2. Map of countries.

We will easily discover the leading research institutes of cross-border e-commerce through institution visual analysis. According to our analysis from Fig. 2, 29 institutions have published relevant articles on cross-border e-commerce. 24 of them from China. Beijing institute of fashion technology is the most productive academic institution. Shanghai jiao tong university has much cooperation with other institutions, such as Beijing union university, chongqing university and wollongong university. In addition, it focused on cross-border e-commerce logistics, platform and consumer satisfaction. Beijing jiao tong university is another influential university. It paid attention to the cross-border e-commerce supply chain, enterprise operation. The cooperation network of analysis result is shown in Fig. 3.

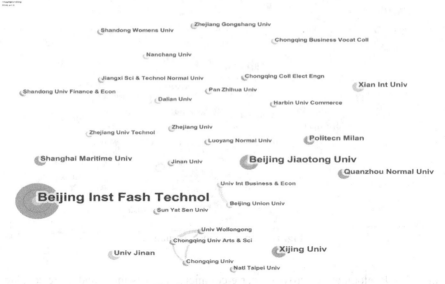

Fig. 3. Map of institutions.

4.3 Mapping and Analysis on Keywords

We have displayed a keyword network in Fig. 3 to show popular research topics. As can be seen, keywords with high centrality include cross-border e-commerce, model, big data, B2B, cross-border e-commerce logistics. Among them, cross-border logistics is a research topic widely discussed. Boyson (Boyson 1999) discoverd that third-party logistics is the main form of cross-border logistics, and how to control and manage third-party logistics is crucial to the development of cross-border e-commerce. Maria Giuffrida (2017) first classified the cross-border e-commerce logistics system in the form of literature review. Based on the above, we can draw a conclusion that cross-border e-commerce consumer behavior and logistics become hot research topic due to the rapid development of wireless Internet technology and huge mobile user market. Through the analysis of co-citation of keywords, the evolution process of cross-border e-commerce can be revealed.

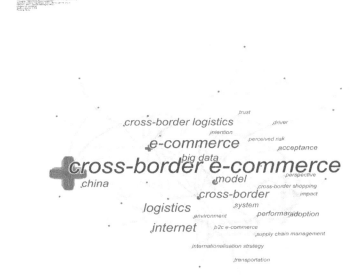

Fig. 4. Map of keywords

4.4 Mapping and Analysis on References

We can discover the highly cited papers and main contribution on cross-border e-commerce through map of references co-citation. In Fig. 4, each node represents one references. The colour of blue represents the earliest years, the colour of green represent mid-range years, the colour of yellow represents late years. Paper from Gomez-herrera E (Gomez-Herrera et al. 2014) is the most cited reference. It focused on drivers and barriers of cross-border e-commerce in the EU. The paper indicated the efficient and flexible cross-border payment systems can increase cross-border e-commerce transactions. The second most frequently co-cited article is from Terzi N, they examine the impact of e-commerce on international trade and employment. It proved countries with high incomes and import volume will benefit from knowledge spillovers. Li et al. (2015) used cultural distance, geographical distance and institutional distance to measure the political uncertainty of cross-border countries and explore how it affects the control level in cross-border acquisition negotiations. Based on the hypothesis of transaction cost economics, logit and random effect regression analysis were used. The empirical results show that cross-border corporations tend to share rather than authorization with the increase of cultural, geographical and institutional distance. In short, eight of the top 10 cited reference adopted the empirical analysis method, one of them is the mathematical model method, and the other was the literature review. The main method of cross-border e-commerce research is empirical analysis. Three of them are about cross-border e-commerce logistics distribution and services, and three of them discussed the constraints of cross-border e-commerce. Cross-border e-commerce logistics and its obstacles are the future research direction (Fig. 5).

Fig. 5. Map of references co-citation

5   Conclusion

We use CiteSpace to analyze the limitations of current research and future research direction. Compared with the existing literature review of cross-border e-commerce, we made a set of knowledge maps to show the future trends of cross-border e-commerce. The result of research enriched bibliometric analysis method and extend types of literature review. These new findings set a clear direction for future research.

(1) The previous researchers paid more attention to cross-border E-commerce platform and consumer behavior. The mainland China published the most relevant articles on cross-border e-commerce. Beijing institute of fashion technology is the most productive academic institution. Shanghai jiao tong university has much cooperation with other institutions. Cross-border logistics have potential become research focus in the future through the above analysis.

(2) Technical problems and soft tools are two constraints in the process. We should promote the integration of mobile payment and cross-border e-commerce (Martens 2013). Identify cross-border e-commerce platform risks under m-commerce environment. More artificial intelligence elements should be introduced into cross-border e-commerce. In terms of the dynamic sustainability, the cross-border e-commerce policy lacked long-term strategic planning, especially in regards to taxation and warehousing policies. The measures for promotion of cross-border e-commerce not only the service system construction and demonstration construction policies, international cooperation and risk monitoring (Liu et al. 2015), but also to conduct comprehensive supervision over all aspects of the cross-border e-commerce supply chain. Thus, a relatively mature system has been established with a particular focus on sustainable development of cross-border e-commerce.

(3) In the context of BRI, improvement of the infrastructure is a key objective of the BRI. The BRI potentially extends he supply chain over a larger geographical distance and through several countries. The BRI links peripheral countries to developed economies providing new markets and new possibilities for sourcing. Challenges in cross border trade specifically for smaller firms intransitional economies also require further research attention. The company should establish a systematic dialogue process with stakeholders to understand and address their expectations. While none of these research issues are particularly new in themselves, the BRI provides a unique context that warrants attention. Further, the BRI context can be used to enhance the research of the theory and practice on the cross-border e-commerce coordination mechanism.

References

1. Ai, W., Yang, J., Lin, W.: Revelation of cross-border logistics performance for the manufacturing industry development. Int. J. Mob. Commun. **14**(6), 593–609 (2016)
2. Boyson, S., Corsi, T.M., Dresner, M.E., Harrington, L.H.: Logistics and the extended enterprise: benchmarks and best practices for the manufacturing professional. Supply Chain Manag. **31**(2), 136–138 (1999)
3. Cao, C., Li, X., Liu, G.: Political Uncertainty and Cross-Border Acquisitions. Social Science Electronic Publishing (2015)
4. Giuffrida, M., Mangiaracina, R., Perego, A., Tumino, A.: Cross-border B2C e-commerce to Greater China and the role of logistics: a literature review. Int. J. Phys. Distrib. Logist. Manag. **47**(6), 00 (2017)
5. Gomez-Herrera, E., Martens, B., Turlea, G.: The drivers and impediments for cross-border e-commerce in the EU. Inf. Econ. Policy **28**(1), 83–96 (2014)
6. Lai, Y., Wang, K.: Cross-border Electronic Commerce's Development Characteristics. Obstacle Factors and the Next Step in China (2014, reform)
7. Liu, X., Chen, D., Cai, J.: The Operation of the Cross-Border e-commerce Logistics in China (2015)
8. Martens, B.: What Does Economic Research Tell Us About Cross-Border E-Commerce in the EU Digital Single Market? Jrc Working Papers on Digital Economy (2013)
9. Synnestvedt, M.B., Chen, C., Holmes, J.H.: CiteSpace II: visualization and knowledge discovery in bibliographic databases. In: AMIA Annual Symposium Proceedings/AMIA Symposium. AMIA Symposium, p. 724 (2005)

Author Index

Printed in the United States
By Bookmasters